A Framework for Priority Arguments

This book presents a unifying framework for using priority arguments to prove theorems in computability. Priority arguments provide the most powerful theorem-proving technique in the field, but most of the applications of this technique are ad hoc, masking the unifying principles used in the proofs. The proposed framework presented isolates many of these unifying combinatorial principles and uses them to give shorter and easier-to-follow proofs of computability-theoretic theorems. Standard theorems of priority levels 1, 2, and 3 are chosen to demonstrate the framework's use, with all proofs following the same pattern. The last section features a new example requiring priority at all finite levels. The book will serve as a resource and reference for researchers in logic and computability, helping them to prove theorems in a shorter and more transparent manner.

Manuel Lerman is a Professor Emeritus of the Department of Mathematics at the University of Connecticut. He is the author of *Degrees of Unsolvability: Local and Global Theory*, has been the managing editor for the book series *Perspectives in Mathematical Logic*, has been an editor of *Bulletin for Symbolic Logic*, and is an editor of the ASL's *Lecture Notes in Logic* series.

LECTURE NOTES IN LOGIC

A Publication of
The Association for Symbolic Logic

This series serves researchers, teachers, and students in the field of symbolic logic, broadly interpreted. The aim of the series is to bring publications to the logic community with the least possible delay and to provide rapid dissemination of the latest research. Scientific quality is the overriding criterion by which submissions are evaluated.

More information can be found at http://www.aslonline.org/books-lnl.html.

A Framework for Priority Arguments

MANUEL LERMAN
University of Connecticut, Emeritus

ASSOCIATION FOR SYMBOLIC LOGIC

CAMBRIDGE
UNIVERSITY PRESS

CAMBRIDGE UNIVERSITY PRESS
Cambridge, New York, Melbourne, Madrid, Cape Town, Singapore,
São Paulo, Delhi, Dubai, Tokyo

Cambridge University Press
32 Avenue of the Americas, New York, NY 10013-2473, USA

www.cambridge.org
Information on this title: www.cambridge.org/9780521119696

Association for Symbolic Logic
Richard Shore, Publisher
Department of Mathematics, Cornell University, Ithaca, NY 14853
http://www.aslonline.org

First published 2010

Printed in the United States of America

A catalog record for this publication is available from the British Library.

Library of Congress Cataloging in Publication data

Lerman, M. (Manuel), 1943–
A framework for priority arguments / Manuel Lerman.
p. cm. – (Lecture notes in logic)
Includes bibliographical references and index.
ISBN 978-0-521-11969-6 (hardback)
1. Proof theory. 2. Priority (Philosophy) 3. Logic, Symbolic and mathematical.
I. Title. II. Series.
QA9.54.L47 2010
511.3′6–dc22 2010005437

ISBN 978-0-521-11969-6 Hardback

CONTENTS

PREFACE

The poset (i.e., partially ordered set) of computably enumerable (i.e., recursively enumerable) degrees $\mathcal{R}$, an algebraic structure that is invariant under the notion of *information content* of sets, was introduced by Post [20] in 1944 as a way to compare the information content of computably enumerable mathematical structures and theories. It seems to have been Post's hope that $\mathcal{R}$ would provide a simple natural hierarchy based on information content, and that, perhaps, this hierarchy would consist of only two comparable degrees. Whether or not this was the case became known as Post's Problem. While this turned out not to be the case, attempts to solve Post's Problem led to an intensive ongoing study of $\mathcal{R}$.

One of the underlying themes of Computability Theory (i.e., Recursion Theory) is to determine the extent to which a given degree structure captures the notion of information content. The hope was that one could find a way to differentiate between sets with different information content within the algebraic structure. The weakest way to do so is to require that the degree structure be *rigid*, i.e., that there be no non-trivial automorphism of the structure. While the rigidity question remains unresolved, there are several results which imply that automorphisms cannot move a given degree too far.

The study of *global* properties of degrees, such as automorphisms, frequently requires an understanding of local structural properties. One way to approach a study of local structure theory is to measure the complexity of a statement as its logical complexity in a fixed language, and to try to find decision procedures for classes of sentences of a given complexity. The historical development of the subject follows an almost monotonic increase in the logical complexity of sentences about $\mathcal{R}$ whose truth is determined. This is because the primary technique used to analyze this structure, the *priority method*, consists of a uniform collection of methods of various levels of complexity, and the level of the method needed to prove that a given sentence is true is closely related to the logical complexity of the sentence. Thus the understanding of increasing levels of the priority method provided greater accessibility to results about $\mathcal{R}$ of higher logical complexity.

The mission of this book is not to present an organized analysis of the current status of knowledge about the elementary theory of $\mathcal{R}$; rather, it is to give a coherent presentation of the priority method. The framework for priority arguments developed by Steffen Lempp and the author is expanded and used to prove sentences about $\mathcal{R}$ of various complexity levels. The sentences, or theorems, chosen represent a rich cross section of results and techniques, and so demonstrate the flexibility of the framework. Furthermore, we are able to separate the technology of the priority method from the analysis necessary to prove a given result. General theorems are proved about the technology of priority arguments. This will allow us to prove theorems by first describing the action taken to satisfy single requirements, and then performing a fairly simple analysis to show that action for specified finite sets of requirements can be compatibly executed.

This monograph is an outgrowth of joint work with Steffen Lempp [10, 11, 12, 13, 14, 15], which began in 1986, and the theorems we prove frequently coincide with those presented (with standard-style proofs) in Lempp's lecture notes [9]. At that time, an attempt was made to prove the decidability of the existential theory of the poset of computably enumerable degrees in a language containing predicates representing the relations $\boldsymbol{a}^{(n)} \leq \boldsymbol{b}^{(n)}$ for all integers n. The approach taken towards the solution of that problem required priority method constructions at all levels of the arithmetical hierarchy. A general framework was developed, over the course of seven years, and used to solve the problem. It should be noted that other general frameworks were simultaneously being developed. The earliest of these was developed by Lachlan [8] in 1970 for lower levels of the arithmetical hierarchy. Particular theorems were proved in the early 1980s using priority methods at an infinite level of the hyperarithmetical hierarchy following an approach developed by Harrington. Shortly after our work began in 1986, Ash [1, 2] developed a general framework for certain types of priority constructions occurring in Computable Model Theory and covering all levels of the hyperarithmetical hierarchy. More general hierarchies of this type have since been developed by Ash and Knight (cf. [6]) but do not seem to be easily applicable to the types of constructions encountered in dealing with $\mathcal{R}$. However, some of the ideas introduced by Ash became an integral part of the framework developed jointly by Lempp and the author. Michalski [18] was able to carry out low-level priority method constructions for $\mathcal{R}$ by generalizing a framework of Ash and Knight. Another framework was under development by Groszek and Slaman, but while the ideas were general, the details were carried out only for low-level priority method constructions.

During the time that Lempp and the author were developing their original framework for the priority method, it became evident that the applicability of the framework extended beyond the theorem that was being proved. Thus in

1993, a study of the extent of applicability of the framework was begun. Initial attempts aimed at complete uniformity, but it was soon discovered that different types of requirements needed to be treated differently, so a modular approach was initiated. This approach attempted to separate the combinatorics of the priority method from the combinatorics needed to satisfy individual requirements. While this approach was successful, it required a separate description of the implementation of the framework for each construction. Furthermore, the proofs of the satisfaction of each new type of requirement made use of properties of the framework, so this approach was unattractive to the uninitiated. Many of the proofs seemed to be ad hoc in nature, and we were not comfortable with the presentation. Thus the project was temporarily suspended in 1995. When it was resumed in 1997, a modified approach that removed some of the original restrictions imposed by uniformity was discovered, thereby providing the flexibility to prove more general theorems about the framework. The approach has several steps. We begin the proof of each theorem by describing the basic module used to satisfy requirements. This module is a finite tree, with a sentence whose truth or falsity directs action at a given non-terminal node of the tree, and sentences describing the action to be taken based on the truth value of the directing sentence. The module is pasted into the highest-level tree, and a level-by-level description of the assignment of sentences to nodes of lower-level trees is presented; the new sentences are obtained from the old ones by bounding certain quantifiers. Instead of requiring the quantifier bounding algorithm to be the same for every theorem, we just list properties that must be obeyed by this algorithm. The construction takes place at the lowest-level tree, representing the computable level. We then verify that the conditions placed on assigning requirements, decomposing sentences, and specifying where the action takes place are satisfied; normally, these conditions can be verified by inspection. We appeal to theorems about the framework to complete the proof. Thus proofs are more transparent, shorter, and much easier to present.

Early in the development of the framework, goals were set for measuring its usefulness. These included:

- The approach should closely resemble the standard approach to priority method constructions using a tree of strategies.
- The approach should cover all priority method constructions used to determine properties of $\mathcal{R}$.
- The method should isolate the combinatorial lemmas which are part of the priority method from those which relate only to a restricted set of results proved by the method.
- The approach should be intuitive and helpful in finding proofs of new results.
- The approach should simplify the process of presenting proofs which can be more easily followed.

We summarize the current status of achievement of the goals, from our point of view.

- The approach uses a "system of trees of strategies" that is close to the standard "tree of strategies" approach. This should make the system of trees of strategies approach as easy to follow as other approaches to general frameworks, for those conversant with the standard approach. However, there is an additional level of abstraction that occurs in passing to a general framework and causes the approach to be less intuitive for beginners in the subject.
- We have undertaken a broad study of theorems proved by priority method constructions, and all seem to be amenable to our methods. A rich cross section of such theorems is presented in this monograph.
- We have already discussed the separation of priority from the analysis of the satisfaction of requirements. The level of separation is greater than that which was originally sought.
- Again, there is an additional level of abstraction that occurs in passing to a general framework and causes the approach to be less intuitive for beginners in the subject. We have used the approach to prove new theorems, and the intuition provided by the approach was of assistance in analyzing conflicts between strategies and finding ways to resolve them.
- We leave it to the reader to determine whether the presentation of the proofs is simpler, a subjective question. Certainly, for some proofs, shortcuts violating basic assumptions needed for the approach to succeed can be taken and used to simplify a construction. For the most part, however, the trees of strategies approach follows the same pattern as the standard approach, and the modularization makes it unnecessary to repeat the parts of the proof for which combinatorial lemmas about the framework are available. There are more combinatorial lemmas than those presented in this book, and we view the development of the framework as an ongoing process.

I would like to acknowledge my debt to Steffen Lempp for the development of the ideas in this monograph. These ideas grew out of earlier joint work with Lempp that would not have been completed without his sharp insight and persistence. His input concerning early drafts of this monograph were invaluable. Thanks are also extended to C. Ash, B. Englert, M. Groszek, L. Harrington, J. Knight, A. Kučera, R. Shore, and T. Slaman for helpful comments related to this work. Finally, I express my love and deep gratitude to my wife Maxine, and children Elliot and Sharon. Without their patience, tolerance, and understanding, this project would never have been completed.

Chapter 1

INTRODUCTION

This chapter is devoted to the presentation of basic definitions and notation to be used in this monograph. The definitions fall under three general headings; those related to computable partial functionals and computably enumerable sets, those related to the computably enumerable degrees, and those related to trees.

1.1. Computably Enumerable Sets

Let $\mathbb{N}$ be the natural numbers, i.e., the set of integers $\{0, 1, 2, \dots\}$. We use interval notation on $\mathbb{N}$; thus $[k, m] = \{n : k \leq n \leq m\}$. Open interval notation and half-open interval notation is used in a similar fashion. The direct sum of two subsets A and B of $\mathbb{N}$ is denoted as $A \oplus B$ and is defined as $\{2x : x \in A\} \cup \{2x + 1 : x \in B\}$. $|A|$ will denote the cardinality of the set A.

If $A \subset \mathbb{N}$, $m \in \mathbb{N}$, and Φ is a partial functional, then we write $\Phi(A; m) \downarrow$ if m is in the domain of $\Phi(A)$, and $\Phi(A; m) \uparrow$ otherwise. If Φ and Ψ are partial functionals and A and B are sets, then we write $\Phi(A) \simeq \Psi(B)$ if $\Phi(A)$ and $\Psi(B)$ are *compatible*, i.e., for all x, if $\Phi(A; x) \downarrow$ and $\Psi(B; x) \downarrow$, then $\Phi(A; x) = \Psi(B; x)$; and we write $\Phi(A) = \Psi(B)$ if $\Phi(A)$ and $\Psi(B)$ are *identical*, i.e., $\Phi(A)$ and $\Psi(B)$ are compatible and for all x, $\Phi(A) \downarrow$ iff $\Psi(B) \downarrow$.

Intuitively, a *computable partial functional* Φ is one for which one can write a computer program using an arbitrary oracle A (i.e., which allows instructions of the form: "If $n \in A$ then go to line r") such that for any $m, k \in \mathbb{N}$, $\Phi(A; m) = k$ if and only if the program produces output k when given input m, and $\Phi(A; x) \uparrow$ if the program fails to halt on input m. A *computably enumerable set* is the domain of a computable partial function $\Phi(\emptyset)$, and B is *computably enumerable in A* if there is a computable partial functional Φ such that B is the domain of $\Phi(A)$. If Φ is a computable partial functional and $\Phi(A; m) \downarrow = k$, then the computer program provides a computation for the *axiom* $\Phi(A; m) \downarrow = k$. A is said to be the *oracle*, m the *argument*, and k the *value* for this axiom (or computation). As the computation halts, only finitely many questions of the form "Is $n \in A$?" are asked; the *use* of the axiom (or computation) is the largest such n.

If an oracle has the form $A \oplus B$, then we write $\Phi(A, B)$ in place of $\Phi(A \oplus B)$. In this case, we allow different uses for A and B in an axiom. If only a single use is specified, then we adopt the convention that this is the common use of A and B. $A \restriction m+1$ is the subset of $[0, m]$ which satisfies $(A \restriction m+1)(k) = A(k)$ for all $k \leq m$. Note that if an axiom has oracle A and use m, then it produces the same computation as would be produced from oracle $A \restriction m+1$; thus we may allow finite sets of the form $A \restriction m+1$ as oracles for axioms.

If Φ is a partial functional with oracle A, then we say that $\lim_s \Phi(A; x, s) = m$ if $\Phi(A; x, s) \downarrow = m$ for all sufficiently large s. We say that $\lim_s \Phi(A)$ is *well-defined* if for all x, either there is an m such that $\lim_s \Phi(A; x, s) = m$, or $\Phi(A; x, s) \uparrow$ for all sufficiently large s.

It was shown by Kleene [5] that there is an effective enumeration $\{\Phi_i : i \in \mathbb{N}\}$ of all computable partial functionals of a fixed number of integer variables; and there is a computable one-to-one function mapping sequences of integers to $\mathbb{N}$. Thus there is no loss of generality in assuming that each such functional has only one integer argument. (We will, however, use functionals with multiple oracles and arguments, and note that the above enumeration uniformly induces effective enumerations of all computable partial functionals with fixed numbers of set and integer variables.) Similarly, there is a computable enumeration $\{W_i : i \in \mathbb{N}\}$ of all computably enumerable sets. Furthermore, there is a computable sequence of approximations to the computable enumeration of functionals or sets, i.e., an array $\{W_i^s : i, s \in \mathbb{N}\}$ such that for all i, $W_i = \cup\{W_i^s : s \in \mathbb{N}\}$, $\max(W_i^s) = s$, and $\{\langle m, s\rangle : m \in W_i^s\}$ is computable. There is also an array $\{\Phi_i^s : i, s \in \mathbb{N}\}$ such that for all i, A and m, $\Phi_i(A; m) = \lim_s \Phi_i^s(A \restriction s+1; m)$; in fact, if $A = W_j$, then $\Phi_i(W_j; m) = \lim_s \Phi_i^s(W_j^s; m)$, and we can assume that $\{\langle i, j, s, \sigma\rangle : \sigma \subset W_j^s \ \&\ \Phi_i^s(\sigma; m)\downarrow\}$ is computable; without loss of generality, we assume that if $\langle i, j, s, \sigma\rangle$ is in this set, then $i, j \leq s$ and $\sigma \subseteq [0, s]$. The notation $\Phi(A; m)[s]$ will be used for $\Phi_i^s(A; m)$.

1.2. Degrees

Define $A \leq_T B$ if there is a computable partial functional Φ such that $\Phi(B) = A$, and $A \equiv_T B$ if $A \leq_T B$ and $B \leq_T A$. $\equiv_T$ is an equivalence relation known as *Turing equivalence* which captures the notion of information content. The equivalence classes, $\{A : A \equiv_T B\}$ are called *degrees*, and form a partially ordered set (*poset*) with ordering $\leq$ induced by $\leq_T$. Lower-case bold-face letters such as $\boldsymbol{a}$ are used to denote degrees. A degree is *computably enumerable* if it contains a computably enumerable set. $\mathcal{R}$ denotes the poset of computably enumerable degrees, and $\boldsymbol{R}$ denotes the set of computably enumerable degrees.

The degrees support a join operation: Thus if $\boldsymbol{a}$ is the degree of A and $\boldsymbol{b}$ is the degree of B, then $\boldsymbol{a} \cup \boldsymbol{b}$ denotes the degree of $A \oplus B$ which is the least upper bound of $\boldsymbol{a}$ and $\boldsymbol{b}$. Some, but not all pairs of degrees have a meet; when such a meet exists, it is denoted by $\boldsymbol{a} \cap \boldsymbol{b}$. We write $\boldsymbol{a}|\boldsymbol{b}$ if $\boldsymbol{a}$ and $\boldsymbol{b}$ are incomparable.

There is a smallest degree, $\boldsymbol{0}$, the degree of the computable sets. The degrees also support a *jump operator* denoted by $\boldsymbol{a}'$; given a set A of degree $\boldsymbol{a}$, $\boldsymbol{a}'$ is the largest degree of a set that is computably enumerable in A. (We remark that such a set always exists.) Thus $\boldsymbol{0}'$ is the largest computably enumerable degree; we fix a computably enumerable set K of degree $\boldsymbol{0}'$.

The *high/low hierarchy* for the computably enumerable degrees is defined as follows. A computably enumerable degree $\mathbf{a}$ is low_n if $\boldsymbol{a}^{(n)} = \boldsymbol{0}^{(n)}$, and is $high_n$ if $\boldsymbol{a}^{(n)} = \boldsymbol{0}^{(n+1)}$. Computably enumerable degrees $\mathbf{a}$ satisfying

$$\boldsymbol{0}^{(m)} < \boldsymbol{a}^{(m)} < \boldsymbol{0}^{(m+1)}$$

for all m are called *intermediate*. A computably enumerable set A is said to be low_n ($high_n$, *intermediate*, resp.) if its degree is low_n ($high_n$ intermediate, resp.).

We recall the Shoenfield Limit Lemma [23]:

LEMMA 1.2.1 (Limit Lemma). *$A \leq_T K$ iff there is a total computable function f such that for all x, $\lim_s f(x,s) = A(x)$.* ⊣

We note that our definition of limit applies to partial functions as well. It is then easy to verify the following lemma, which we will use as our version of the Limit Lemma:

LEMMA 1.2.2 (Limit Lemma). *$A \leq_T K$ iff there is a computable partial function f such that for all x, $\lim_s f(x,s)\downarrow = A(x)$.*

1.3. Finite Sequences and Trees

Let A be a computable set of symbols, and let $T(A) = A^{<\omega}$, i.e., the set of all finite sequences of symbols from A. $\emptyset$ will denote the sequence of length 0. For $\sigma \in T(A)$, let $|\sigma|$ denote the length of the sequence σ, i.e., the cardinality of the domain of σ. If there is an $n < \min\{|\sigma|, |\tau|\}$ such that $\sigma(n) \neq \tau(n)$, then we write $\sigma|\tau$. For $\sigma, \tau \in T(A)$, we define $\sigma^\frown\tau$ to be the sequence of length $|\sigma| + |\tau|$ consisting of the sequence of symbols σ followed by the sequence of symbols τ.

We define a partial order, $\subseteq$, on $T(A)$ by $\sigma \subseteq \rho$ if there is a τ such that $\sigma^\frown\tau = \rho$. $\langle T, \subseteq\rangle$ is a *tree* if T is an initial segment of $T(A)$ for some A, and $\subseteq$ is the restriction of the ordering on $T(A)$ to T. If $\sigma, \tau \in T(A)$, then $\sigma \wedge \tau$ is the longest $\rho \in T(A)$ such that $\rho \subseteq \sigma$ and $\rho \subseteq \tau$; and if $|\sigma| > 0$, then σ^-

is the unique $\rho \subset \sigma$ such that $|\sigma| = 1 + |\rho|$. We also use interval notation for trees; thus $[\sigma, \tau] = \{\rho : \sigma \subseteq \rho \subseteq \tau\}$. Similar notation is used for open and half-open intervals.

A *path* through a tree $T = T(A)$ is an ω-sequence of elements of A; $[T]$ denotes the set of all paths through T. (More generally, *trees* are defined as posets having the property that the set of predecessors of any element is linearly ordered, and a *path* through a tree is a maximal linearly ordered subset of the tree.)

Sequences of trees will be specified; the notation used is $\{T^i : i \leq n\}$. In this context, an element of T^i will be of the form σ^i, i.e., the superscript will identify the tree of the sequence on which σ^i lies. If $i = 0$, then the superscript will be omitted.

We will be using paths through one tree to approximate to paths through another tree through a limit approximation, which is defined as follows. Fix $\tau \in T \cup [T]$ and a function λ from T to another tree T'. Then $\xi \subseteq \lim\{\lambda(\eta) : \eta \subset \tau\}$ iff $\xi \subset \lambda(\eta)$ for a cofinal set of nodes $\eta \subseteq \tau$.

Chapter 2

SYSTEMS OF TREES OF STRATEGIES

Many theorems concerning structural properties of $\mathcal{R}$ are proved by constructing sets satisfying each requirement in an infinite list. Frequently, the action taken to satisfy one requirement will conflict with the preservation of action taken earlier to satisfy another requirement. The priority method was devised to organize the action taken to satisfy requirements so that, at the end of a construction, all requirements are satisfied.

As the logical complexity (measured by the complexity of sentences in a certain language) of the requirements under consideration increases, it becomes more difficult to analyze the conflicts between requirements and to show that all requirements are satisfied. The *systems of trees of strategies* approach is an inductive method which provides such an analysis. The proofs that we present using this approach all follow the same pattern.

We begin with a list of all requirements to be satisfied, and it will usually be clear that the theorem will follow once we show that all requirements are satisfied. We will then introduce *basic modules* for each requirement; these modules will describe, in a high-level language, the manner in which the requirement will be satisfied. The basic modules will be finite binary trees whose edges and non-terminal vertices or *nodes* are labeled with sentences. Each non-terminal node will have a *directing sentence* whose role is to direct the action taken; if the directing sentence is true, then we will follow the instructions of the *validated action* sentence along one edge emanating from that node, and if the sentence is false, then we will follow the instructions of the *activated action* sentence along the other edge emanating from that node. The basic modules will be templates, and the sentences will be allowed to use parameters defined in terms of nodes of the basic module, and will be allowed to restrict quantifiers involving these nodes to lie along the true path through the tree. There will then be instructions on how to implement the basic module on the highest-level tree of strategies T^n, and an observation that the implementation faithfully reflects the basic module and the requirement.

The next step will be the decomposition of requirements, level by level, from T^n to the computable level T^0 at which the construction takes place. Suppose that we have determined what takes place on T^{k+1}, and wish to pass to T^k.

Each node $\eta^{k+1} \in T^{k+1}$ will have potentially infinitely many *derivatives* on T^k; these will be nodes working to ensure that the instructions for η^{k+1} are carried out. Thus we will need an algorithm that determines the node of T^{k+1} for which a given node of T^k is working, and a definition of the path through T^{k+1} computed, through a limit approximation, by a given path through T^k. The trees of strategies will need to be defined carefully in order to reflect this computation. We will then need to describe how the directing sentence and action sentences for η^{k+1} are decomposed to produce the sentences for the derivative $\eta^k \in T^k$ of η^{k+1}. This will usually be done by bounding some quantifiers, and we will again allow nodes of T^k to be used as parameters in the bounding process. We will have to show that this process is a faithful reflection of the requirement, namely, that if we follow the instructions of the sentences for the derivatives of η^{k+1} along a path Λ^k through T^k and Λ^k computes the path Λ^{k+1} through T^{k+1} along which η^{k+1} lies, then we will have satisfied the instructions of the sentences for η^{k+1}.

When we reach T^0, then we will have instructions for an effective construction. However, we will need to show that these instructions can be implemented. Thus we cannot have instructions telling us to withhhold a number x from a computably enumerable set A if x has already been placed into A, nor can we have instructions to declare an axiom $\Phi^s(\sigma; x) = m$ when we have already declared an axiom $\Phi^t(\tau; x) = k$ for some $t < s$, $\tau \subseteq \sigma$ and $k \neq m$. Similarly, we cannot have such conflicting instructions at a given stage s. Once this is shown, we implement the construction and the theorem then follows. There may be nodes along the true path through T^0 whose instructions we choose to ignore, but in all such cases, we will show that these instructions are derived from a node that does not lie on the true path through its tree.

As is evident from the above description, there are many things that need to be checked or proved for a given theorem. If separate proofs were required for each theorem, there would be no advantage to this approach, but this is not the case. Many of the steps to be carried out will be the implementation of one of several fixed algorithms used over and over, and we will have theorems showing that if an algorithm has certain easily checked properties, then it accomplishes its task. Thus the proof of each theorem will consist of checking, usually by inspection, that certain properties hold, and quoting some theorems about the framework. These theorems will apply to many constructions, and thus prevent duplication of work in proofs. Thus once we have the needed theorems describing properties of the framework, the hard part in proving a theorem will consist of designing basic modules, and sometimes revising fixed algorithms, so that the necessary properties are satisfied.

In Section 2.1, we present an overview of this chapter, and use the proof of the Friedberg–Mučnik Theorem to motivate the trees of strategies approach. Trees of strategies are introduced in Section 2.2. The properties to be obeyed by basic modules are presented in Section 2.3. The λ *function*, defined in Section 2.4, will provide a computable limit approximation of a path Λ^{k+1} through T^{k+1} from a path Λ^k through T^k. The *links* that are defined in Section 2.5 will be used to keep track of nodes $\eta^{k+1} \in T^{k+1}$ that are free to be *switched* by λ, i.e., nodes for which it is safe to have $\lambda(\eta^k)|\lambda((\eta^k)^-)$ with $\eta^{k+1} = \lambda(\eta^k) \wedge \lambda((\eta^k)^-)$. In Section 2.6, we define the nodes that are eligible to be *antiderivatives* of η^k, i.e., from which η^k is allowed to be *derived*. Each construction will have to describe the way the antiderivative of η^k is chosen from among these eligible nodes. Antiderivatives for nodes of T^k will be chosen within *blocks* of nodes, i.e., segments of T^k containing no infinite paths. Properties of the block formation process are specified in Section 2.7, and the assignment of requirements to nodes of trees and the decomposition of requirements will also be discussed in that section. The *weight function*, introduced in Section 2.8, will supply bounds for the quantifiers and will determine arguments for functionals.

Directing sentences and action are discussed in Sections 2.9 and 2.10. The first of these sections deals primarily with an example, and the second section with more general properties of these sentences. The Framework Theorem is presented in Section 2.11. This section does not require a deep understanding of the framework, and can be read without understanding many of the details of prior sections. We present two simple proofs of lemmas about the framework in Section 2.12.

2.1. An Overview

This chapter is devoted to the development of the trees of strategies framework. Much of this development is inductive in nature, and terminology may be encountered that has not yet been defined. One of the purposes of this section is to give an overview of the development of the framework, and in particular, to introduce the concepts and terminology used and the intuition behind the concepts. In order to make the intuition more concrete, we will interleave a description of the proof of the Friedberg–Mučnik Theorem within the framework. We begin with an introduction to this theorem.

The study of $\mathcal{R}$ traces its history back to Post's [20] fundamental paper of 1944. Post focused on the computably enumerable sets, sets that can be enumerated by a computer, and began the study of the information content of such sets by trying to determine properties of $\mathcal{R}$. There are two special computably enumerable degrees, $\boldsymbol{0}$, the degree of the computable sets which is the smallest degree in this poset, and $\boldsymbol{0}'$, the largest computably enumerable degree. Post asked whether there were other computably enumerable degrees, a question that became known as *Post's Problem*.

Solutions to Post's Problem were found independently by Friedberg [4] and Mučnik [19] more than a decade later through the construction of a pair of computably enumerable sets whose degrees are incomparable. These solutions introduced a new technique, the *priority method*, that is the subject of this monograph. As the proof of this result is one of the simplest applications of the priority method, it will be used to motivate the systems of trees of strategies approach.

THEOREM 2.1.1 (Friedberg, Mučnik). *There are computably enumerable sets A and B such that $A \not\leq_T B$ and $B \not\leq_T A$.*

Each construction will have an associated level, and a level n construction will take place on the sequence of trees $T^0, \dots, T^n$. The description of the construction will begin on T^n, and will descend, one level at a time, until we reach the computable level T^0. The levels go hand-in-hand with the arithmetical hierarchy, as for all k, an oracle of degree $\boldsymbol{0}^{(k)}$ will be able to determine the action of the construction on T^k. For each $k \leq n$, a portion of the construction will be assigned to each node η^k of T^k, and if $k > 0$, then η^k will divide its assignment among many nodes of T^{k-1}. Furthermore, each node of T^{k-1} will receive an assignment from at most one node of T^k. If η^k delegates part of its assignment to $\eta^{k-1} \in T^{k-1}$, then we will call η^{k-1} a *derivative* of η^k, and will call η^k the *antiderivative* of η^{k-1}. We will also introduce notation to track this relationship; we say that $\text{up}(\eta^{k-1}) = \eta^k$ if η^k is the antiderivative of η^{k-1}.

A typical construction will have to satisfy each requirement in a given sequence of requirements. In a level n construction, each requirement will have a level, and that level will be $\leq n$.

EXAMPLE 2.1.2. In the proof of the Friedberg–Mučnik Theorem, we will construct two computably enumerable sets, A and B. The construction will aim to satisfy the following requirements, for each computable partial functional Φ.

$$P_\Phi : \Phi(A) \neq B.$$

$$Q_\Phi : \Phi(B) \neq A.$$

The manner in which requirements are to be satisfied will be presented through a labeled finite binary tree called a *basic module*. A *directing sentence* (in prenex normal form) will be assigned to each non-terminal node of the basic module, and each such node will have a level. If the level of the node is k, then the directing sentence will be a Σ^0_k sentence if k is odd, and a Π^0_k sentence if k is even. *Action sentences* will be assigned to each of the two edges emanating from a non-terminal node. Suppose that we are given two successive nodes, α and β, of a basic module, and that $\beta = \alpha^\frown\langle\gamma\rangle$; then we call γ an *outcome* of α. If the directing sentence assigned to α is Σ^0_k (Π^0_k, resp.), then we refer to γ as a Σ-*outcome* (Π-*outcome*, resp.) and say that α has

Σ *outcome along* β (Π *outcome along* β, resp.) if β guesses that the directing sentence is true, and we refer to γ as a Π-*outcome* (Σ-*outcome*, resp.) and say that α has Π *outcome along* β (Σ *outcome along* β, resp.) if β guesses that this directing sentence is false. We also want terminology that keeps track of the truth values of directing sentences in a level-independent manner. Thus we refer to γ as a *validated outcome* (*activated outcome*, resp.) if β guesses that the directing sentence assigned to α is true (false, resp.), and say that α is *validated along* β (*activated along* β, resp.). Note that the identification of the activated vs. validated designation with the Σ vs. Π designation alternates, level by level. We will later describe how the nodes of the basic module are identified with nodes of T^n.

EXAMPLE 2.1.3. We describe the basic module for the requirement $\Phi(A) \neq B$ of the Friedberg–Mučnik Theorem. The module will consist of a single non-terminal node α having level 1, with two terminal successors α_0 and α_1. α will have activated and Π outcome along α_0, and will have validated and Σ outcome along α_1. The directing sentence will be

$$\exists u \exists s (\Phi(A \upharpoonright u; x)[s] = 0)$$

where x will be a parameter determined by the framework from the node of T^1 to which α is assigned. The activated action sentence will be

$$\forall t (x \notin B^t),$$

and the validated action sentence will be

$$\exists r \forall t \geq s (x \in B^r \ \& \ A^t \upharpoonright u = A^s \upharpoonright u).$$

(In actuality, we will restrict the range of the variable s, but we need not be concerned with that aspect here.)

While a computable construction is to be carried out along T^0, the action sentences are Π^0_1 sentences. (In Example 2.1.3 the action sentences presented are those for T^1; when we pass to T^0, parameters will have been chosen to replace r, s and u, and the quantifier $\exists r$ will disappear.) These are to be interpreted as instructions to the construction at later stages. Thus we will carry out the instructions of the action sentences for all t that are less than or equal to the weight of nodes ρ^0 along the true path through T^0 which extend the node $\eta^0 \in T^0$ to which the requirement is assigned. We will not necessarily implement the action at all such nodes ρ^0, but it will suffice for us to implement action at infinitely many such nodes, as long as it seems that η^0 is a node for which the construction requires action to be carried out.

Let $\eta^{k-1} \in T^{k-1}$ be a derivative of $\eta^k \in T^k$, and suppose that S is the directing sentence assigned to η^k. For concreteness, we assume that k is odd; a similar pattern is observed when k is even. We observe from the above comments that S is a Σ^0_k sentence. If S is not properly Σ^0_k (i.e., if S is Π^0_{k-1}

or Σ^0_{k-1}), then S is also the directing sentence assigned to η^{k-1}. Otherwise, the directing sentence $\widetilde{S}$ for η^{k-1} is obtained by bounding each quantifier in the block of leading existential quantifiers in S; thus $\widetilde{S}$ will be a Π^0_{k-1} sentence. (This is why validated outcomes switch from Σ outcomes on T^k to Π outcomes on T^{k-1}.) The particular bounds are frequently generated by the *weight function* wt, a function whose domain consists of the nodes of the trees of strategies, and which, on each tree, is monotonically increasing with the length of the node.

EXAMPLE 2.1.4. We again analyze the sentences used in the proof of the Friedberg–Mučnik Theorem. Suppose that the directing sentence of Example 2.1.3 is assigned to $\eta^1 \in T^1$, and let $\eta^0 \in T^0$ be a derivative of η^1. The directing sentence for η^0 will have the form

$$\exists u \leq p \exists s \leq q(\Phi(A \restriction u; x)[s] = 0),$$

where p and q are number parameters generated by the weight function evaluated on η^0 and nodes generated from η^0. The activated action sentence for η^0 will be

$$\forall t \leq w(x \notin B^t),$$

and the validated action sentence for η^0 will be

$$\forall t \in [s, w](x \in B^r \ \& \ A^t \restriction u = A^s \restriction u),$$

where r and w are other parameters generated by the nodes that are currently being visited. The parameter p is the weight of the longest node on T^1 lying along the current path through T^1 computed by η^0, the parameter q is the weight of η^0 itself, the parameter r will be the weight of the successor node to η^0 on the path determined by the construction, and the parameter w will be the weight of the current path through T^0. The reason for these choices will become apparent when we see what is needed to make the proof succeed; for the time being, it is only important to note the bounding of quantifiers.

For $k \in (0, n]$, the correspondence between the outcome of a node $\eta^k \in T^k$ along a path $\Lambda^k \in [T^k]$ and the outcomes of its derivatives along a path $\Lambda^{k-1} \in [T^{k-1}]$ which generates Λ^k must preserve predicted truth values of directing sentences. We make this concept more precise with the consideration of the form of directing sentences. Suppose that the directing sentence S for η^k is a Σ^0_k sentence (a similar analysis can be carried out for Π^0_k sentences).

First suppose that Λ^k predicts that S is true, i.e., that S is validated along Λ^k. Then Λ^k predicts that we will find a witness for each quantifier in the leading block of existential quantifiers of S. This prediction must be borne out along Λ^{k-1}; thus there must be a derivative η^{k-1} of η^k along Λ^{k-1} at which these witnesses are found. This, in turn, requires the bounds that η^{k-1} places on the quantifiers just mentioned to be sufficiently large to bound the witnesses we seek, and so the directing sentence $\widetilde{S}$ for η^{k-1} must be true. As $\widetilde{S}$ is a Π^0_{k-1} sentence, η^{k-1} must have Π outcome along Λ^{k-1}. Furthermore, there will be no need to have derivatives of η^k along Λ^{k-1} that extend the Π outcome of η^{k-1}, as we have already verified the predicted truth value for η^k.

Now suppose that Λ^k predicts that S is false, i.e., that S is activated along Λ^k. Then Λ^k predicts that we will never simultaneously find witnesses for all of the quantifiers in the leading block of existential quantifiers of S. This prediction must be borne out along Λ^{k-1}; thus there must be infinitely many derivatives η^{k-1} of η^k along Λ^{k-1} at which we try, and fail, to find these witnesses, and the bounds on the above-mentioned quantifiers placed at such nodes η^{k-1} must be arbitrarily large. In this case, the directing sentences for such η^{k-1} must be false. As the latter sentences are Π^0_{k-1} sentences, each such η^{k-1} must have Σ outcome along Λ^{k-1}.

We will code outcomes in a way that tracks these correspondences. Thus if the directing sentence S for η^k is predicted to be true along its immediate successor $\rho^k \subset \Lambda^k$, then the outcome of ρ^k along Λ^k will be the validated immediate successor $\nu^{k-1} \subset \Lambda^{k-1}$ of the node η^{k-1} at which the witnesses to the truth of S were found. In this way, we code not only the fact that the node η^k was validated, but also where the witnesses were found, enabling us to continue the process of determining where it was, on T^{k-2}, that η^{k-1} was determined to be validated. If S is predicted to be false along Λ^k, then there is no single node along Λ^{k-1} that made the prediction. Thus we set the outcome of ρ^k along Λ^k to be the activated immediate successor ν^{k-1} of the shortest derivative of η^k encountered along Λ^{k-1}. This outcome makes the prediction that we will fail to succeed at finding witnesses at all derivatives of η^k along Λ^{k-1}. In both cases, we have specified that ν^{k-1} is the outcome of ρ^k along Λ^k; we introduce notation to denote this fact, writing $\text{out}(\rho^k) = \nu^{k-1}$. We will call η^{k-1} the *principal derivative* of η^k along Λ^{k-1} (and also along all nodes of T^{k-1} extending ν^{k-1}). We will also call the shortest derivative μ^{k-1} of η^k along Λ^{k-1} (and also along all proper extensions of μ^{k-1} along Λ^{k-1}) the *initial derivative* of η^k along Λ^{k-1}. (Note that outcomes denoting action for η^k are outcomes of ρ^k rather than of η^k, as the outcome of the action for η^k along a path is only determined at its immediate successor along that path.)

If the initial derivative μ^{k-1} of η^k along Λ^{k-1} differs from the principal derivative π^{k-1} of η^k along Λ^{k-1}, then all action prescribed for nodes in the interval $[\mu^{k-1}, \pi^{k-1})$ is based on an incorrect prediction of the truth value of η^k, as long as the current path through T^{k-1} extends the Π outcome of π^{k-1}. Thus we wish to halt all action taken for nodes captured in this interval. We track this by forming a primary Λ^{k-1}-*link* $[\mu^{k-1}, \pi^{k-1}]$, and preventing all nodes except π^{k-1} captured by this link from supplying derivatives to trees of lower level than $k-1$ while the prediction witnessed at the validated immediate successor ν^{k-1} of π^{k-1} seems valid. Many of the nodes in the interval will be derivatives of nodes of higher-level trees that do not lie on the current path; but nodes $\subset \eta^k$ may have derivatives captured in this link, and their antiderivatives may lie on the current path. These nodes must also be prevented from supplying derivatives to lower level trees. Thus we say that the nodes in $[\mu^{k-1}, \pi^{k-1})$ are Λ^{k-1}-*restrained.* Nodes σ^j along the current path ρ^j through a tree T^j with the property that for all $i \in [j, n]$, the antiderivative of σ^j on T^i lies along the current path through T^i as computed by ρ^j and is not restrained along that path are said to be *free* at ρ^j.

The above description leads to the definition of the trees of strategies. As directing sentences on T^0 are quantifier-free, the outcomes on T^0 will be taken from the set $\{0, \infty\}$; 0 will be the prediction that the directing sentence is false, and ∞ will be the prediction that the directing sentence is true. For $k > 0$, T^k will be the set of finite increasing sequences of nodes of T^{k-1}. Such a sequence will code the principal derivative sequence of an initial segment of nodes along a path through T^k. The computation of the path Λ^k computed by Λ^{k-1} will be generated by a *path generating function* λ. For $\rho^{k-1} \in T^{k-1}$, $\lambda(\rho^k)$ will compute by starting with the shortest node of T^k and finding its principal derivative along ρ^{k-1}. This will produce the next node along the path, and $\lambda(\rho^{k-1})$ will determine its outcome as its principal derivative along ρ^{k-1}. This process will continue until we reach a node of T^k that has no derivatives $\subset \rho^{k-1}$, and will stop at that point with its computation of the *current path* through T^k at ρ^{k-1}.

We now describe the process used to determine the correspondence between requirements and nodes of trees, and so to define the derivative process. We begin on T^n, the tree of greatest level for the particular construction, with an assignment of requirements. In most cases, there will be a standard way of doing so, namely by effectively listing all requirements in an ω-sequence and assigning the jth requirement to all nodes of T^n of length j. But in cases where the basic module describing action taken for the requirement is not of minimal length, we will have to exercise a bit more care. We will impose easily checked conditions covering all constructions, that will suffice for the proofs of the corresponding theorem.

We now proceed by descent, going from T^{k+1} to the tree of next highest level, T^k. On that tree, we proceed by induction on the length of nodes; the shortest node of T^k will be a derivative of the shortest node of T^{k+1}. The inductive process looks at immediate successors η^k of nodes whose antiderivatives have already been determined, and specifies an antiderivative for η^k. The node η^k determines an outcome for $(\eta^k)^-$, and so we can compute a new current path at η^k; the definition of the path generating function λ is presented in Section 2.4. Once we have a current path, we can determine the links along that path, and so, the nodes that are restrained along that path; this is carried out in Section 2.5. We must now specify the nodes of T^{k+1} that are eligible to be antiderivatives of η^k. We have already described the restrictions to be placed to make this determination. The antiderivatives of all such nodes must lie on the current paths through their respective trees as determined by η^k, and must be free along those paths; if a node along $\lambda(\eta^k)$ has a derivative with Π outcome along η^k, we have completed the determination of the truth value of its directing sentence, so it is ineligible to be the antiderivative of η^k; finally, a node $\sigma^{k+1} \subseteq \lambda(\eta^k)$ that has level $\leq k$ may only supply one derivative along any path through T^k, so may not be the antiderivative of η^k if it already has a derivative $\subset \eta^k$. These conditions are the *consistency conditions* delineated in Section 2.6. The antiderivative assignment process is described in Section 2.7. There is a standard process that can be used in many cases, but other cases require more care. Thus rather than specifying an assignment process, we specify conditions governing assignment processes which suffice to ensure that the constructions will be successful. At this point, we will only note that the process is carried out in *blocks*; this provides enough control to ensure that nodes requiring infinitely many derivatives along a path will get them without interfering with the ability to assign derivatives to nodes which do not yet have an initial derivative along η^k.

The success of the *iterated trees of strategies* approach for a level n construction relies on having faithful communication between trees. First of all, the assignment of requirements to paths through T^n must ensure that if we faithfully carry out the instructions assigned to each node along the path, then the theorem will follow. Next, for $k < n$, we must ensure that the instructions passed down from a node η^{k+1} that is free along a path $\Lambda^{k+1} \in [T^{k+1}]$ are faithfully executed by the free derivatives of η^{k+1} along a path $\Lambda^k \in [T^k]$ that computes Λ^{k+1}. Finally, we must ensure that the construction faithfully carries out the instructions of all free nodes lying along the true path $\Lambda^0 \in [T^0]$. The verification of these conditions seems to be essentially parallel to the verification needed in standard constructions on a single tree. However, we will be able to prove lemmas covering many theorems stating that if the processes described above satisfy certain easily verifiable conditions, then the faithfulness required will follow immediately. It is this latter ability which we believe will be of value in simplifying the work necessary to present new proofs.

We conclude this section with an overview of the proof of the Friedberg–Mučnik Theorem.

EXAMPLE 2.1.5. We begin by forming a computable list of order-type ω of all requirements listed in Example 2.1.2, and assign the *kth* such requirement to all nodes of T^1 of length k. The sentences associated with each requirement are described in Example 2.1.3, and the parameter x is chosen to be the weight of the node of T^1 to which the requirement is assigned. The choice of antiderivatives for nodes $\eta^0 \in T^0$ is the standard one; we choose the shortest consistent node of T^1 that has no derivative in the block containing η^0 as the antiderivative of η^0, and the sentences assigned to that node are described in Example 2.1.4. The construction now follows the instructions assigned to nodes of T^0, following a path reflecting the truth values of the directing sentences assigned to those nodes. Action sentences for η^0 are applied to all nodes along the true path Λ^0 which extend η^0, as long as η^0 is free at the given node; once η^0 ceases to be free, it will never again be free, and all its action sentences are ignored.

The only part of the proof not covered by noting that all goes well as certain easily verified properties are obeyed by this description, is that which shows that there is no conflict between positive action sentences for η^0 (sentences placing a number into A or B), and negative action sentences for η^0 (instructions to withhold numbers from A or B). But this will follow easily from a lemma stating that if the positive action sentences to place x into a set (say the set A) always choose x to be the weight of the antiderivative η^1 of η^0 on T^1 and are only implemented for validated action, and negative action sentences to withhold x from A are always imposed by nodes whose weight is $\geq x$, then there is no conflict between the positive and negative action. This is a lemma that can be applied to many constructions, and having been proved once, allows one to verify that a construction is successful by checking only easily verifiable conditions.

2.2. Systems of Trees

Requirements and subrequirements will reside on trees, and their action will be coordinated through the use of a *system of trees*. The action taken by the construction at a node of the tree T^0 will be determined by the truth or falsity of the directing sentence associated with that node. Thus T^0 is chosen to be a binary tree. Nodes of T^0 will be finite sequences from $\{0, \infty\}$, and *outcomes* of the action taken at a node will be designated as Σ *outcomes* (*finite outcomes*) or Π *outcomes* (*infinite outcomes*). For $k > 0$, the outcomes of nodes of T^k will be nodes of T^{k-1}. The outcome ρ^{k-1} of a node $\eta^k \in T^k$ will be used to determine how subrequirements assigned to nodes derived from η^k were satisfied on all trees T^j for $j < k$. ρ^{k-1} will be an immediate successor of a node η^{k-1} derived from η^k. Examples of this use of outcomes will be given after the trees are defined.

T^k will be defined as the set of finite increasing sequences of nodes from T^{k-1}. These are the only paths through the set of all finite sequences of nodes from T^{k-1} that are accessible through the approximation process.

DEFINITION 2.2.1. We set $T^0 = \{0, \infty\}^{<\omega}$. If $0 < k \leq n$ and T^{k-1} has been defined, let

$$T^k = \{\eta^k \in (T^{k-1})^{<\omega} : \forall i, j < |\eta^k|(i < j \rightarrow \eta^k(i) \subset \eta^k(j))\}.$$

$\mathcal{T}^k = \langle T^k, \subseteq\rangle$ is the *tree of strategies of level* k, ordered by inclusion. We refer to the elements of T^k as *nodes* of T^k. If $i \in \{0, \infty\}$ and $\rho = \eta^\frown\langle i\rangle \in T^0$, then we call i the *outcome* for η along ρ. Suppose that $k > 0$. Each successor node of T^k is identified with its immediate predecessor followed by a node of T^{k-1}. If $\rho^k \in T^k$, $\eta^{k-1} \in T^{k-1}$, and $\rho^k = (\rho^k)^{-\frown}\langle\eta^{k-1}\rangle$, then η^{k-1} is the *outcome for* $(\rho^k)^-$ *along* ρ^k. A function is introduced to compute outcomes of nodes; in the above situations, we define $\text{out}(\rho^k) = i$ if $k = 0$, and $\text{out}(\rho^k) = \eta^{k-1}$ if $k > 0$. If $j \leq k$, then we define $\text{out}^j(\rho^k)$ by reverse induction; $\text{out}^k(\rho^k) = \rho^k$, and $\text{out}^{j-1}(\rho^k) = \text{out}(\text{out}^j(\rho^k))$. (It might be less confusing to write $\text{out}((\rho^k)^-, \rho^k)$ rather than $\text{out}(\rho^k)$ for the outcome of $(\rho^k)^-$ along ρ^k; but as $(\rho^k)^-$ is uniquely determined by ρ^k, we have chosen to economize on notation.)

The following condition follows from the requirement that nodes on T^{k+1} be increasing sequences of nodes on T^k:

$$(k > 0 \;\&\; \eta^k \subset \rho^k \in T^k) \rightarrow \text{out}(\eta^k) \subset \text{out}(\rho^k).$$

There are two types of outcomes, activated and validated, and the type of outcome followed depends on whether the directing sentence is false or true. As blocks of like unbounded quantifiers are bounded in passing from level to level, the leading block of unbounded quantifiers for the directing sentences at consecutive levels alternates between a block of universal quantifiers and a block of existential quantifiers. For a sentence with a leading block of universal (existential, resp.) quantifiers, the Π outcome (Σ outcome, resp.) is the one that guesses that the sentence is true, and the Σ outcome (Π outcome, resp.) is the one that guesses that the sentence is false, i.e., that its negation is true. Thus the correspondence between activated and validated outcomes on the one hand, and Σ outcomes and Π outcome on the other hand, will alternate level by level.

DEFINITION 2.2.2. Fix $k \leq n$, $\eta^k \in T^k$ and $\rho^k \in T^k \cup [T^k]$ such that $\rho^k \supseteq \eta^k$ and $|\eta^k| > 0$. If $k = 0$, then we say that $(\eta^k)^-$ is *activated* (*validated*, resp.) *along* ρ^k if $\text{out}(\eta^k) = 0$ ($\text{out}(\eta^k) = \infty$, resp.). If $k > 0$, then $(\eta^k)^-$ is *activated* (*validated*, resp.) *along* ρ^k if $(\text{out}(\eta^k))^-$ is activated (validated, resp.) along $\text{out}(\eta^k)$. We say that $(\eta^k)^-$ has Σ *outcome* (Π *outcome*, resp.) *along* ρ^k if either $k = 0$ and $\text{out}(\eta^k) = 0$ ($\text{out}(\eta^k) = \infty$, resp.), or $k > 0$ and $(\text{out}(\eta^k))^-$ has Π outcome (Σ outcome, resp.) along $\text{out}(\eta^k)$. (Note that $(\eta^k)^-$ is activated (validated, resp.) along η^k if either k is even and $(\eta^k)^-$ has Σ outcome (Π outcome, resp.) along η^k, or k is odd and $(\eta^k)^-$ has Π outcome (Σ outcome, resp.) along η^k.)

In order to provide the reader with some intuition about systems of trees, a proof using these systems is now related to the single *tree of strategies* approach introduced by Harrington and popularized by Soare [25], and the methods of treatment of certain concrete requirements by these approaches are compared. Recall the Friedberg–Mučnik requirements discussed in Section 2.1. These are discussed again in the context of systems of trees. Much of this discussion duplicates the earlier discussion, but the new terminology and context allows for more precision.

EXAMPLE 2.2.3. Consider a typical Friedberg–Mučnik requirement $\Phi(A) \neq B$. The standard tree of strategies approach assigns such a requirement to a node η^1 of $T^1 = \{0,1\}^{<w}$, and proceeds by stages. When η^1 first appears on the true path, a *follower* x is assigned to the requirement. As long as η^1 is on the path through T^1 computed at stage s and $\Phi(A;x)[s] \neq 0$, the path computation at s follows $\eta^{1\frown}\langle 0\rangle$, and the construction sets $B^{s+1}(x) = 0$. (η^1 is *activated* along this path.) If, at some t greater than or equal to the stage at which η^1 first appears on the true path computation, we find that $\Phi(A \restriction u; x)[t]\downarrow = 0$ for some $u \leq t$, then the path computation at t follows $\eta^{1\frown}\langle 1\rangle$, the construction sets $B^{t+1}(x) = 1$, $A \restriction u$ is restrained, and η^1 never acts again. (η^1 is *validated* along this path.)

The systems of trees of strategies approach replaces stages with nodes of the tree $T^0 = \{0,\infty\}^{<w}$. In place of stage t, the *weight* (see Definition 2.8.2) of the node $\eta^0 \in T^0$ derived from η^1 at which the construction has arrived, is used. Let $\rho^0 = \eta^{0\frown}\langle\beta\rangle$ be the immediate extension of η^0 along Λ^0; $\rho^1 = \eta^{1\frown}\langle\rho^0\rangle \in T^1$ is used to code whether or not $\Phi(A;x)\downarrow = 0$. The outcome ρ^0 of η^1 identifies whether or not the requirement has been activated or validated, and in addition, that the decision to activate or validate was made based on the outcome of η^0 along ρ^0. If η^1 is activated (so has Π outcome) along ρ^1, then η^1 will have infinitely many derivatives along the true path $\Lambda^0 \in T^0$, all of which will be activated (and have Σ outcome) along Λ^0. The immediate successor $\rho^1 = \eta^{1\frown}\langle\eta^{0\frown}\langle 0\rangle\rangle$ of η^1 will identify η^0 as the derivative of η^1 at which the decision to determine the outcome of η^1 along ρ^1 is made, namely, the first derivative of η^1 along Λ^0 (η^0 is designated both as the *initial* and *principal* derivative of η^1 along Λ^0), and the outcome 0 of η^0 along Λ^0 indicates that η^0 is activated along Λ^0. If η^1 is validated (so has Σ outcome) along ρ^1, then the immediate successor $\rho^1 = \eta^{1\frown}\langle\eta^{0\frown}\langle\infty\rangle\rangle$ of η^1 will identify the node η^0 of T^0 at which the decision to determine the outcome of η^1 along ρ^1 was made (η^0 is designated as the *principal derivative* of η^1 along ρ^1), and the outcome ∞ of η^0 along Λ^0 indicates that η^0 is validated (so has Π outcome) along ρ^0.

If η^0 is not the first (i.e., initial) derivative μ^0 of η^1 along ρ^1 and η^0 has Π outcome along ρ^0, then we create a link from μ^0 to η^0. Links partially correspond, in standard priority arguments, to initializing all extensions of ρ^1. At higher levels, they also serve the purpose of preventing nodes captured by the link from acting to change the approximation to the true path. This will ensure that when the outcome of a node is switched by the approximation, the switch must be caused by action taken for a derivative of the node that is switched.

EXAMPLE 2.2.4. Consider a typical *thickness* requirement on T^2. An infinite computable set R is given, activation corresponds to allowing only finitely many elements of R to enter a set A, and validation corresponds to forcing cofinitely many elements of R to enter A. Suppose that this thickness requirement is assigned to a node η^2 along the true path $\Lambda^2 \in [T^2]$. Then η^2 will have derivatives along the true path $\Lambda^1 \in [T^1]$, each of which will have the role of placing elements of R into A if its associated Σ_1 sentence is true.

First suppose that one of these sentences is false, say the one corresponding to the derivative η^1 of η^2. Then η^1 will be the last (and *principal*) derivative of η^2 along Λ^1, and will have Π outcome along Λ^1, indicating that no derivative of η^1 is validated along the true path $\Lambda^0 \in [T^0]$, i.e., that no derivative of η^1 finds a witness for its existential sentence at the stage corresponding to its weight. As there will be infinitely many derivatives of η^1 along Λ^0, we will be verifying the falsity of the directing sentence for both η^1 and η^2. No elements greater than or equal to the least element of R for which η^1 has responsibility are placed into A in this case.

Now suppose that all directing sentences for derivatives of η^2 along Λ^1 are true. Then η^2 will have Π outcome along Λ^2, indicating that η^2 has infinitely many derivatives along Λ^1, each of which is validated (so has Σ outcome) along Λ^1, i.e., has a derivative along Λ^0 whose directing sentence is true. Each such node will place the elements of R for which it is responsible into A. All but finitely many elements of R will be assigned to such nodes, so cofinitely many elements of R will be placed into A. (The Π outcome of η^2 along Λ^2 is the *initial* derivative η^1 of η^2 along Λ^1 (which is also the *principal* derivative of η^2 along Λ^1), followed by its first validated derivative η^0 along Λ^0 and the outcome ∞ for η^0 indicating that η^0 is validated.)

2.3. Basic Modules

A basic module is a finite labeled binary directed tree that is used to describe the strategy for satisfying a given requirement. Two edges will exit from each non-terminal node; one will be used to describe activated action, and the other to describe validated action. In addition, the edges will determine Σ and Π outcomes for the node, and the correspondence between the Σ/Π, and activated/validated categorizations will be consistent with that of Definition 2.2.2 when nodes of the basic module are assigned to trees. Each non-terminal node of the basic module will have an associated *directing sentence*, and each edge will be assigned an *action*; action will be categorized as *activated* or *validated* depending on the label given to the edge. Each node will have a *level*, determined by the complexity of its directing sentences and action sentences. Usually, the level will be the complexity of the node's directing sentence; thus a node having a Σ_k^0 directing sentence will have level k.

Nodes of a basic module will be assigned to nodes of the highest-level tree T^n. Terminal nodes will be used only to determine paths for activation or validation, and will not be assigned directing sentences. Thus when basic modules are assigned to trees, the terminal nodes of the basic module will be identified with initial nodes of a new basic module. The conditions required of basic modules are now axiomatized; they will be easy to verify by inspection for a given construction.

DEFINITION 2.3.1 (Basic Modules). A basic module is a labeled directed tree $\langle T, \leq \rangle$ with the following properties:

(i) T is finite, and $|T| > 1$.
(ii) T has a unique initial element.
(iii) Every non-terminal element of T has exactly two immediate successors.

The non-terminal nodes of T will each have a *level*, and will be assigned a *directing sentence*. One edge of T will be classified as *activated* and the other will be classified as *validated*, and in addition, one edge of T will be classified as Σ and the other will be classified as Π. *Activated action* will be assigned to each activated edge of T, and *validated action* will be assigned to each validated edge of T. No directing sentence is assigned to any terminal node of T.

REMARK 2.3.2. We use lower-case Greek letters, generally from the beginning of the alphabet, to represent nodes of basic modules, and these nodes will have subscripts from $\{0, 1\}^{<\omega}$. The ordering of these nodes will correspond to the lexicographical ordering of the subscripts, and the empty sequence will produce no subscript. If σ is such a subscript, then $\alpha_{\sigma\frown\langle 0\rangle}$ will represent the immediate successor (in the tree ordering of the basic module) of α_σ along which the activated outcome is followed, and $\alpha_{\sigma\frown\langle 1\rangle}$ will represent the immediate successor (in the tree ordering) of α_σ along which the validated outcome is followed.

2.4. The Path Generating Function λ

The next four sections will constitute a simultaneous induction, first on $n-k$ and then on $|\eta^{n-k}|$ for $\eta^{n-k} \in T^{n-k}$. During this induction, we will need to define how requirements are assigned to nodes of trees, the up function connecting nodes of different trees working for the same requirement, the path computation function λ, links, and the notions of *initial derivative* and *principal derivative*. For the purposes of the induction, we will assume that we are at a node $\eta^k \in T^k$, and that the functions and concepts in question have been defined for T^j if $j > k$ and also for $\xi^k \subset \eta^k$. We begin with the definition of λ. This function is not defined for nodes of T^n.

If $k < n$, then each node $\eta^k \in T^k$ will be derived from a node $\eta^{k+1} \in T^{k+1}$. In this case, η^k is said to be a *derivative* of η^{k+1}, written as $\mathrm{up}(\eta^k) = \eta^{k+1}$. The path generating function λ provides an approximation to an initial segment $\lambda(\eta^k)$ of a path through T^{k+1}. If $\Lambda^k \in [T^k]$, then $\lim\{\lambda(\eta^k) : \eta^k \subset \Lambda^k\}$ will be a path through T^{k+1}.

The definition of the path generating function λ is meant to capture the following situation. Each $\xi = \xi^k \subset \eta^k$ will be derived from a node $\sigma = \sigma^{k+1} \in T^{k+1}$. A directing sentence S_σ will be associated with σ, and will give rise to a directing sentence S_ξ for ξ. Suppose that S_σ begins with a universal quantifier. (This assumption implies that the activated outcome coincides with the Σ outcome and the validated outcome coincides with the Π outcome at the associated level.) If σ has level $\geq k+1$, we will frequently obtain S_ξ by bounding the leading block of universal quantifiers in S_σ by a parameter. As long as each S_ξ is true, the approximation given by λ predicts that $\sigma^{k+1\frown}\langle\mu^{k\frown}\langle\rho^{k-1}\rangle\rangle \subseteq \lambda(\eta^k)$, where μ^k is the initial derivative of σ^{k+1} along η^k (defined below) and ρ^{k-1} is the (validated and Σ) outcome of μ^k along η^k. If we find a first $\xi = \xi^k$ for which S_ξ is false, then $\sigma^{k+1\frown}\langle\xi^{k\frown}\langle\rho^{k-1}\rangle\rangle \subseteq \lambda(\eta^k)$, where ρ^{k-1} is the (activated and Π) outcome of ξ^k along η^k; in this case, ξ^k will be the principal derivative of σ^{k+1} along η^k. If S_σ begins with an existential quantifier, then we proceed as above after replacing S_ξ with $\neg S_\xi$, and interchanging *activated* and *validated*. (If $\mathrm{lev}(\sigma^{k+1}) \leq k$, then the outcome of σ^{k+1} along $\lambda(\eta^k)$ is induced by the outcome of the unique derivative of σ^{k+1} along η^k.)

We formally define $\lambda(\eta^k)$ as follows:

DEFINITION 2.4.1. Suppose that $\mathrm{up}(\xi^k)$ is defined for all $\xi^k \subset \eta^k$. If $\eta^k \neq \langle\,\rangle$, let $\nu^k = (\eta^k)^-$. It will follow inductively from (2.6.1) and Definition 2.7.7(i) that $\mathrm{up}(\eta^k) \subseteq \lambda((\eta^k)^-)$. We then define $\lambda(\eta^k)$ as follows:

$$\lambda(\eta^k) = \begin{cases} \langle\,\rangle & \text{if } \eta^k = \langle\,\rangle, \\ \mathrm{up}(\nu^k)^\frown\langle\eta^k\rangle & \text{if } \eta^k \neq \langle\,\rangle \text{ and either } \mathrm{up}(\nu^k) = \lambda(\nu^k), \text{ or } \nu^k \text{ is} \\ & \text{the shortest } \xi^k \subset \eta^k \text{ such that both } \mathrm{up}(\xi^k) = \\ & \mathrm{up}(\nu^k) \text{ and } \xi^k \text{ has } \Pi \text{ outcome along } \eta^k, \\ \lambda((\eta^k)^-) & \text{otherwise.} \end{cases}$$

(The requirement, in the second clause of the definition of $\lambda(\eta^k)$, that ν^k be as short as possible will be superfluous, as condition (2.6.2) below will be added later and will imply that there is at most one ξ^k satisfying the requirement of the second clause.) For $i \in [k, n]$, $\lambda^i(\eta^k)$ is defined as follows: $\lambda^k(\eta^k) = \eta^k$ and $\lambda^i(\eta^k) = \lambda(\lambda^{i-1}(\eta^k))$ for $i > k$. And if $\Lambda^k \in [T^k]$, then

$\lambda(\Lambda^k) = \lim_s\{\lambda(\Lambda^k \upharpoonright s)\}$, and $\Lambda^{k+1} = \lambda(\Lambda^k)$. For all $i \in [k, n]$, $\lambda^i(\eta^k)$ is called the *current path through T^i at η^k*.

REMARK 2.4.2. It follows from Definition 2.4.1 that:

(i) If $|\sigma^{k+1}| > 0$ and $\sigma^{k+1} \subseteq \lambda(\eta^k)$, then $\text{out}(\sigma^{k+1}) \subseteq \eta^k$.
(ii) If $\lambda((\eta^k)^-) \supseteq \sigma^{k+1}$ and $\lambda(\eta^k) \not\supseteq \sigma^{k+1}$, then for all $\delta^k \supseteq \eta^k$, $\lambda(\delta^k) \not\supseteq \sigma^{k+1}$.
(iii) If $\lambda(\eta^k) \supseteq \sigma^{k+1}$, then σ^{k+1} has a derivative $\subset \eta^k$ iff $\lambda(\eta^k) \supset \sigma^{k+1}$.
(iv) If $\xi^k \subset \eta^k \in T^k$, $\lambda(\xi^k)|\lambda(\eta^k)$, and $\kappa^{k+1} = \lambda(\xi^k) \wedge \lambda(\eta^k)$, then there is a derivative $\widetilde{\kappa}^k \subset \xi^k$ of κ^{k+1} that has Σ outcome along ξ^k (in fact, all derivatives of κ^{k+1} that are $\subset \xi^k$ have Σ outcome along ξ^k), and a derivative $\kappa^k \in [\xi^k, \eta^k)$ of κ^{k+1} that has Π outcome along η^k.
(v) If $\xi^k \subset \eta^k$ and $\lambda(\xi^k)$ and $\lambda(\eta^k)$ are comparable, then $\lambda(\xi^k) \subseteq \lambda(\eta^k)$.
(vi) If $\sigma^{k+1} \subset \lambda(\eta^k)$, then σ^{k+1} has Π outcome along $\lambda(\eta^k)$ iff all derivatives of σ^{k+1} have Σ outcome along η^k.

(The verification of (ii) follows by first noting that this condition follows straightforwardly for $\kappa^{k+1} = \lambda((\eta^k)^-) \wedge \lambda(\eta^k)$ in place of σ^{k+1}, and once we see that (ii) holds for κ^{k+1}, (ii) follows easily for $\sigma^{k+1} \supseteq \kappa^{k+1}$.)

EXAMPLE 2.4.3. Suppose that the sequence $\eta_5 = \langle 0, 0, 0, 0, \infty \rangle \in T^0$ is given. For all $i \leq 5$, let $\eta_i = \eta_5 \upharpoonright i$. Let $\tau_0 = \emptyset \in T^1$, and suppose that η_0, η_1 and η_3 are derivatives of τ_0. Then $\lambda(\eta_0) = \tau_0$, and $\lambda(\eta_1) = \lambda(\eta_2) = \tau_0{}^\frown\langle \eta_0{}^\frown\langle 0\rangle\rangle = \langle\langle 0\rangle\rangle = \tau_1$. Now suppose that η_2 and η_4 are derivatives of τ_1. Then $\lambda(\eta_3) = \lambda(\eta_4) = \tau_1{}^\frown\langle \eta_2{}^\frown\langle 0\rangle\rangle = \tau_1{}^\frown\langle \eta_3\rangle = \tau_2$. Finally, $\lambda(\eta_5) = \tau_1{}^\frown\langle \eta_4{}^\frown\langle \infty\rangle\rangle = \tau_1{}^\frown\langle \eta_5\rangle = \tau_3$. Note that τ_2 and τ_3 are incomparable.

Two derivatives of $\xi^{k+1} \in T^{k+1}$ along η^k play special roles. The first of these is the initial derivative of ξ^{k+1} along η^k, defined in Definition 2.7.6, which is the shortest derivative of ξ^{k+1} along η^k. The second is the principal derivative of ξ^{k+1} along η^k, defined below. Principal derivatives of nodes are the derivatives that determine outcomes.

DEFINITION 2.4.4. Fix $\pi^k \subset \eta^k \subseteq \rho^k \in T^k \cup [T^k]$ and $\sigma^{k+1} \in T^{k+1}$ such that $(\eta^k)^- = \pi^k$. We say that π^k is the *principal derivative* of σ^{k+1} *along* ρ^k if $\text{up}(\pi^k) = \sigma^{k+1}$, and either π^k has Π outcome along ρ^k and is the shortest derivative of σ^{k+1} that has Π outcome along ρ^k, (it will follow from (2.6.3) and Definition 2.7.7 that at most one derivative of σ^{k+1} has Π outcome along ρ^k), or all derivatives of σ^{k+1} along ρ^k have Σ outcome along ρ^k and π^k is the initial derivative of σ^{k+1} along ρ^k. (Thus we do not require that $\text{up}(\pi^k) \subseteq \lambda(\rho^k)$.) If $i \geq k$ and $\sigma^i \in T^i$, we call π^k the *principal derivative* of σ^i *along* ρ^k if either $i = k$ and $\pi^k = \sigma^i$, or $i > k$ and there is a $\pi^{k+1} \in T^{k+1}$ such that π^k is the principal derivative of π^{k+1} along ρ^k and π^{k+1} is the principal derivative of σ^i along $\lambda(\rho^k)$. For $j \leq k+1$, we define $\text{prin}^j(\sigma^{k+1}, \rho^j)$ to be the principal derivative of σ^{k+1} along ρ^j, if it exists; when $j = 0$, we will frequently omit the superscript j.

REMARK 2.4.5. It follows from Remark 2.4.2(iv) and Definition 2.4.4 that if $\pi_0^k \subset \pi_1^k$, $\lambda(\pi_0^k) \mid \lambda(\pi_1^k)$, and $\kappa^{k+1} = \lambda(\pi_0^k) \wedge \lambda(\pi_1^k)$, then there are $\mu^k \subset \pi_0^k \subseteq \pi^k \subset \pi_1^k$ such that μ^k is the principal derivative of κ^{k+1} along π_0^k and has Σ outcome along π_0^k so is the initial derivative of κ^{k+1} along π_0^k and π^k is the principal derivative of κ^{k+1} along π_1^k and has Π outcome along π_1^k.

2.5. Links

Suppose that $k < i \leq n$. We will need to place restrictions on the nodes of T^i that are are eligible to be *switched* by a given node along the current path through T^k. One restriction requires a node to be *free* when it is switched, i.e., that it not be contained in any *link*. Links are formed when a switch occurs, and can be broken when the outcome of a switched node is switched back. Thus in Example 2.4.3, the node η_5 switches τ_1.

Links correspond to initialization, after injury, in the standard approach to infinite injury priority arguments. Suppose that a node $\sigma^2 \in T^2$ has shortest derivative μ^1 along a path Λ^1 through T^1, and principal derivative $\pi^1 \supset \mu^1$ along $\rho^1 \subset \Lambda^1$. Then we form a primary ρ^1-link $[\mu^1, \pi^1]$ from μ^1 to π^1, thereby restraining any node $\xi^1 \in [\mu^1, \pi^1)$ from acting and destroying axioms declared by π^1. (Note that if $[\mu^1, \pi^1]$ is a ρ^1-link, then π^1 is not restrained by $[\mu^1, \pi^1]$. However, as we can have $[\mu^1, \pi^1) = [\mu^1, \delta^1)$ as intervals with $\pi^1 \neq \delta^1$, we use closed interval notation $[\mu^1, \pi^1]$ for ρ^1-links to make sure that there is a one-to-one correspondence between intervals that determine links, and the links themselves.) Any $\xi^1 \in [\mu^1, \pi^1)$ will either be a derivative of a node η^2 that is no longer on the current path, or a derivative of a node $\eta^2 \subseteq \sigma^2$. Links on T^j are also used to prevent action by the derivatives of η^2 within the interval that they determine, as such nodes will either be acting based on an incorrect guess to the current path through some tree, or will switch the current path on T^k for some $k > j$ at a node that is not an antiderivative of a node that is switched on T^i for some $i \in [j, k]$. In other words, this restriction on action will allow a node $\sigma^k \in T^k$ to be switched only when a derivative of σ^k receives a new Π outcome. If η^2 is never returned to the current path of T^2, then η^2 will have no derivatives extending π^1, and will not need to have its requirement satisfied. If $\eta^2 \subseteq \sigma^2$, then it will be shown that there is no harm in preventing derivatives of η^2 restrained by $[\mu^1, \pi^1]$ from acting, as any such action can be assigned to derivatives of η^2 extending π^1. Derivatives of π^1 will be allowed to act, as π^1 is not restrained by $[\mu^1, \pi^1]$.

DEFINITION 2.5.1. A node $\eta^k \in T^k$ such that $|\eta^k| > 0$ is said to be *switching* if there is an $i > k$ such that $\lambda^i((\eta^k)^-) | \lambda^i(\eta^k)$. For the least such i, we say that η^k is *i-switching*. If $j \in [i, n]$ and η^k is i-switching, we say that η^k *switches*

$\lambda^j((\eta^k)^-) \wedge \lambda^j(\eta^k)$. (We will show in Lemma 8.3.3 (Link Analysis) that in this case, $\lambda^j((\eta^k)^-) \wedge \lambda^j(\eta^k) = \mathrm{up}^j((\eta^k)^-)$.)

Fix $\rho^k \in T^k$. Each ρ^k-link will be derived from a *primary* $\lambda^j(\rho^k)$-link for some $j \geq k$, and will have either Σ *outcome* or Π *outcome*. The ρ^k-*links* of T^k will be defined by induction on $n - k$. If $k = n$, then there are no ρ^k-links. Suppose that $k < n$.

DEFINITION 2.5.2. Suppose that $\mu^k \subset \pi^k \subset \rho^k \in T^k \cup [T^k]$, π^k is the principal derivative of $\pi^{k+1} = \mathrm{up}(\pi^k)$ along ρ^k and μ^k is the shortest derivative of π^{k+1} along ρ^k. Then $[\mu^k, \pi^k]$ is a *primary* ρ^k-*link* and has Π *outcome*.

EXAMPLE 2.5.3. Consider the discussion of the Friedberg–Mučnik Theorem in Example 2.2.3. There are no links on T^1. A link $[\mu^0, \pi^0]$ on T^0 restrains those nodes that acted for derivatives of nodes on T^1 between the time $\mathrm{up}(\mu^0) = \mathrm{up}(\pi^0)$ was initialized at μ^0, and the time that the need to realize the follower of $\mathrm{up}(\mu^0)$ was observed at π^0. Such action is no longer viable. No meaningful action was taken for nodes $\subset \mathrm{up}(\mu^0)$, and the path extending $\mathrm{up}(\mu^0)$ is switched, corresponding to initialization of requirements assigned to nodes properly extending $\mathrm{up}(\mu^0)$.

As indicated above, nodes $\xi^k \in T^k$ that are contained in primary links must be prevented from acting. As action takes place on T^0, the fact that ξ^k is contained in a primary link must be transmitted to T^0. The role of *derived links*, defined below, is to transmit this information by pulling down ρ^k-links to T^{k-1}. In actuality, the definition pulls down $\lambda(\rho^k)$-links on T^{k+1} to derived ρ^k-links on T^k.

DEFINITION 2.5.4. Fix $k \leq n$ and $\rho^k \in T^k \cup [T^k]$. Suppose that $[\mu^{k+1}, \pi^{k+1}]$ is a $\lambda(\rho^k)$-link. Assume that the initial derivative μ^k of μ^{k+1} along ρ^k and the principal derivative π^k of π^{k+1} along ρ^k both exist. Then $[\mu^k, \pi^k]$ is a ρ^k-*link derived from* $[\mu^{k+1}, \pi^{k+1}]$. $[\mu^k, \pi^k]$ *has* Σ *outcome* if $[\mu^{k+1}, \pi^{k+1}]$ has Π outcome, and *has* Π *outcome* otherwise. If $[\mu^{k+1}, \pi^{k+1}]$ is derived from a link $[\mu^i, \pi^i]$ for some $i \geq k+1$, then every link derived from $[\mu^{k+1}, \pi^{k+1}]$ is *derived from* $[\mu^i, \pi^i]$.

EXAMPLE 2.5.5. Consider the satisfaction of thickness requirements discussed in Example 2.2.4. There are no links on T^2. A link $[\mu^1, \pi^1]$ on T^1 restrains those nodes that acted for derivatives of nodes on T^2 between the time $\mathrm{up}(\mu^1) = \mathrm{up}(\pi^1)$ was initialized at its initial derivative $\mu^0 \subset \Lambda^0$, and the time that it was observed at the initial derivative π^0 of π^1 along Λ^0 that the directing sentence for π^1 is false. The path extending $\mathrm{up}(\mu^1)$ is switched, corresponding to initialization of requirements assigned to nodes properly extending $\mathrm{up}(\mu^1)$. When a derivative σ^1 of a node $\sigma^2 \subset \mathrm{up}(\mu^1)$ is Λ^1-restrained by $[\mu^1, \pi^1]$, a new derivative of σ^2 is appointed just beyond the link and assigned the responsibility to carry out the action previously assigned to σ^1. This node can only be Λ^1-restrained by links derived from derivatives of nodes $\subset \text{up}(\mu^1)$, so this responsibility will eventually be assigned to a free derivative of σ^2 along Λ^1. No other nodes can be restrained by $[\mu^1, \pi^1]$. $[\mu^0, \pi^0]$ is the Λ^0-link derived from $[\mu^1, \pi^1]$, and invalidates all action taken for nodes between μ^0 and π^0. Such nodes will either be derivatives of nodes that are Λ^1-restrained by $[\mu^1, \pi^1]$, or derivatives of nodes on T^2 that extend a validated outcome of $\text{up}(\mu^1)$ and so do not lie along Λ^2. If π^1 is eventually switched, say by ρ^0, then $[\mu^1, \pi^1]$ will not be a $\lambda(\rho^0)$-link. Furthermore, even though the outcome of π^0 is not switched, $[\mu^0, \pi^0]$ will not be a ρ^0-link as the derivation of this link will no longer be valid.

The *free* nodes are now defined to be those nodes that are not restrained by links. We will only allow free nodes to be switched, and may impose additional restrictions later on the nodes that we allow to be switched.

DEFINITION 2.5.6. Fix $k \leq n$ and $\eta^k \subset \rho^k \in T^k \cup [T^k]$. We say that η^k is *ρ^k-restrained* if there is a ρ^k-link $[\mu^k, \pi^k]$ such that $\mu^k \subseteq \eta^k \subset \pi^k$. In this case, we say that η^k is *ρ^k-restrained by* $[\mu^k, \pi^k]$. If, in addition, $\mu^k \subset \eta^k$, then we say that η^k is *properly ρ^k-restrained (by $[\mu^k, \pi^k]$)*. η^k is *ρ^k-free* if η^k is not ρ^k-restrained.

2.6. Consistency

Only certain nodes of T^{k+1} are eligible to be antiderivatives of the node $\eta^k \in T^k$ (i.e., to be the node $\text{up}(\eta^k)$). The restrictions imposed are those needed to prove lemmas that are applicable to a broad range of priority constructions. Some of the conditions are technical, but seem to be necessary to prove the lemmas.

Nodes σ^{k+1} that are eligible to be *antiderivatives* of η^k must satisfy four conditions. The first condition, (2.6.1), requires the antiderivative of σ^{k+1} on T^i to be on the current path through T^i along the approximation to the path through T^i computed by η^k for each $i \in [k+1, n]$. The second condition, (2.6.2), requires that there be no witness $\xi^k \subset \eta^k$ for the existential sentence associated with σ^{k+1}. In this case, η^k has all the information needed to correctly predict the outcome of σ^{k+1}. The search for witnesses only takes place when $k < \text{lev}(\sigma^{k+1})$, as we cannot begin the process of bounding quantifiers in directing sentences or action sentences before this point. Thus if $k \geq \text{lev}(\sigma^{k+1})$, then we will require σ^{k+1} to have at most one derivative along any path through T^k, and so impose condition (2.6.3). Condition (2.6.4) requires σ^{k+1} to be $\lambda(\eta^k)$-free. This condition will be shown later to imply condition (2.6.1), but for now, it is convenient to require both conditions.

DEFINITION 2.6.1. Fix $\eta^k \in T^k$. We say that σ^{k+1} is η^k*-consistent* if the following conditions hold:

$$\forall i \in [k+1, n](\mathrm{up}^i(\sigma^{k+1}) \subseteq \lambda^i(\eta^k)). \tag{2.6.1}$$

$$(\sigma^{k+1} \subset \lambda(\eta^k) \ \&\ \mathrm{lev}(\sigma^{k+1}) > k) \rightarrow \\ \forall \nu^k \subset \eta^k(\mathrm{up}(\nu^k) = \sigma^{k+1} \rightarrow \nu^k \text{ has } \Sigma \text{ outcome along } \eta^k). \tag{2.6.2}$$

$$\mathrm{lev}(\sigma^{k+1}) \leq k \rightarrow \forall \nu^k \subset \eta^k(\mathrm{up}(\nu^k) \neq \sigma^{k+1}). \tag{2.6.3}$$

$$\sigma^{k+1} \text{ is } \lambda(\eta^k)\text{-free.} \tag{2.6.4}$$

2.7. Blocks

Requirements will be assigned to nodes of trees in *blocks*. $\mathrm{up}(\eta^k) \in T^{k+1}$ is determined when $k < n$ and a requirement is assigned to $\eta^k \in T^k$, and the requirement that is assigned to η^k is the one that was assigned to $\mathrm{up}(\eta^k)$. The assignment process will differ from construction to construction, as care must sometimes be taken to allow all requirements to be satisfied. However, there are some overall conditions on the assignment process that can be axiomatized, and that are amenable to proving general lemmas about all constructions. Furthermore, when the standard assignment is followed on T^k for all $k < n$, then we will be able to separate the tree of strategies analysis from the analysis of the satisfaction of requirements.

The assignment of requirements to nodes begins on T^n. Each block of T^n will correspond to a wide version of a basic module. The width is necessary, as the block must implement the basic module, yet T^n will be infinite branching (a node may have infinitely many activated and validated successors), while a basic module is a finite binary tree.

DEFINITION 2.7.1 (Initial Assignment). Fix $\delta^n \subseteq \tau^n \subseteq \rho^n \in T^n$. If either $\delta^n = \emptyset$, or a δ^n-path through a block of T^n is completed at $(\delta^n)^-$, then δ^n *begins the* δ^n*-block*. Each construction will choose a requirement for the node δ^n that begins the δ^n-block, or for a Δ_n construction, it may be the case that no requirement is assigned to δ^n, in which case δ^n is a terminal node of the tree. If a basic module is assigned to δ^n, then a homomorphism from the block to the basic module for the requirement is defined as follows, and all nodes lying in the same block as δ^n along a path through T^n are said to *lie in the same implementation of a basic module*. We first map δ^n to the initial node of the selected basic module (Definition 2.3.1(ii) ensures the existence and uniqueness of this node). Suppose that $(\rho^n)^-$ has been mapped to a non-terminal node α of the basic module, (in this case $(\rho^n)^-$ is called an α*-node*), and that $(\rho^n)^-$ has activated (validated, resp.) outcome along ρ^n. As α is non-terminal, there is a node β of the selected basic module such that $\beta^- = \alpha$

and α has activated (validated, resp.) outcome along β. We now map ρ^n to β (the well-definedness of this mapping follows from Definition 2.3.1(iii) and the labeling in Definition 2.3.1). A ρ^n-path through the block is *completed at* $(\tau^n)^-$ if τ^n has been mapped to a terminal node α of the basic module; a given construction may also identify other nodes of the basic module as nodes completing blocks along one of their immediate successors. We say that σ^n *is in the* δ^n*-block along* ρ^n if $\delta^n \subseteq \sigma^n \subset \rho^n$, δ^n begins a ρ^n-block, and no $\kappa^n \subset \sigma^n$ completes a ρ^n-path through the δ^n-block. (Note that if a terminal node of a basic module is assigned to ρ^n, then an initial node of a new module may also be assigned to ρ^n. As terminal nodes of basic modules do not have directing sentences, this will not create conflicts.)

The initial assignment of requirements to nodes of T^n will be required to satisfy the Initial Assignment Property, which stipulates that the satisfaction of all requirements along any path through T^n ensures the validity of the corresponding theorem. This property will normally be verifiable by inspection for a given theorem. While we state it here, it will be verified separately within the proof of each theorem, and our framework theorems will conclude only that all requirements are satisfied.

DEFINITION 2.7.2. The initial assignment of requirements to nodes of T^n must satisfy the following *Initial Assignment Property*:

(i) If $\Lambda^n \in [T^n]$ and all requirements assigned to nodes $\subset \Lambda^n$ are satisfied, then the corresponding theorem is true.

Frequently, a standard procedure will be used to assign requirements to nodes of T^n. This procedure is now described, and it is routine to verify that it satisfies the Initial Assignment Property.

DEFINITION 2.7.3 (Standard Initial Assignment). Definition 2.7.1 spells out all aspects of the initial assignment procedure, except for the choice of requirement to be assigned to a node δ^n that begins a block. The *standard initial assignment* chooses the first requirement R (in a fixed computable list of requirements) that has not yet been assigned to any node $\xi^n \subset \delta^n$, and assigns R to δ^n.

LEMMA 2.7.4 (Standard Initial Assignment Lemma). *Suppose that the satisfaction of all requirements in our list suffices to prove a given theorem. Then the standard initial assignment ensures the satisfaction of Definition* 2.7.2(*i*).

Suppose that $k < n$. Nodes completing blocks on T^k are initial derivatives of nodes completing blocks on T^{k+1}.

DEFINITION 2.7.5. Fix $\delta^k \in T^k$ such that either $\delta^k = \emptyset$ or $(\delta^k)^-$ completes a δ^k-block. Then δ^k begins the δ^k*-block*. Fix $\sigma^k \in T^k$. An η^k-path through the δ^k*-block is completed at* $(\eta^k)^-$ if $(\eta^k)^-$ is the initial derivative of up$((\eta^k)^-)$

along η^k, no η^k-path through the δ^k-block is completed at any $\xi^k \subset (\eta^k)^-$ and a $\lambda(\eta^k)$-path through a block of T^{k+1} is completed at $\text{up}((\eta^k)^-)$. In addition, if we have a Δ_n construction, $k = n - 1$, and no requirement is assigned to $\lambda(\delta^k)$, then every derivative ξ^k of the longest node $\xi^n \subset \lambda(\delta^k)$ that has Π outcome along $\lambda(\delta^k)$ *completes a* ρ^k*-block* whenever ξ^k has Σ outcome along ρ^k.

Derivatives and antiderivatives are defined as follows:

DEFINITION 2.7.6. Fix $\eta^k \in T^k$. When the assignment process defines $\text{up}(\eta^k)$, η^k is called a *derivative* of $\text{up}(\eta^k)$. The derivative operation can be iterated; thus for every $j > k$ and $\eta^j \in T^j$ such that $\text{up}(\eta^k)$ is a derivative of η^j, η^k is a *derivative* of η^j and $\text{up}^j(\eta^k) = \eta^j$; η^k is also a *derivative* of η^k. If there is no $\xi^k \subset \eta^k$ such that $\text{up}(\xi^k) = \text{up}(\eta^k)$, then for all $\rho^k \in T^k \cup [T^k]$ such that $\rho^k \supseteq \eta^k$, η^k is the *initial derivative of* $\text{up}(\eta^k)$ *along* ρ^k; and if $j > k$ and $\text{up}(\eta^k)$ is an initial derivative of η^j along $\text{up}(\eta^k)$, then η^k is also an *initial derivative of* η^j *along* every $\rho^k \in T^k \cup [T^k]$ such that $\rho^k \supseteq \eta^k$. η^j is an *antiderivative* of η^k if η^k is a derivative of η^j. For $j \leq k+1$, we define $\text{init}^j(\sigma^{k+1}, \rho^j)$ to be the initial derivative of σ^{k+1} along ρ^j, if it exists; when $j = 0$, we will frequently omit the superscript j.

If $k < n$, the process of determining antiderivatives for nodes of T^k will proceed by induction on $n - k$, perhaps in conjunction with other inductions. For most constructions, a standard assignment procedure will be followed. The framework can tolerate a great deal of additional flexibility for this assignment process, so we will not restrict constructions to the standard procedure alone. Instead, a list of the properties that a derived assignment procedure must satisfy is specified in the next definition. Definition 2.7.7(i) implements the consistency conditions of Section 2.6. Definition 2.7.7(ii) helps to ensure that there are sufficiently many free derivatives along the true path for the construction. And Definition 2.7.7(iii) will be used to show that all blocks have finite intersection with any path through T^k. These clauses will normally be verifiable by inspection for a given construction.

DEFINITION 2.7.7 (Derived Assignment Properties). Suppose that $\eta^k \subset \rho^k \in T^k$, $(\rho^k)^- = \eta^k$, and η^k is in the δ^k-block. If there are any η^k-consistent nodes, then $\text{up}(\eta^k)$ must be defined. In addition, the following *Derived Assignment Properties* must hold:

(i) $\text{up}(\eta^k)$ is η^k-consistent.
(ii) If a ρ^k-path through the δ^k-block is completed at η^k, then each ρ^k-consistent $\eta^{k+1} \subset \lambda(\rho^k)$ has a derivative in the δ^k-block.
(iii) If a block is begun at $\delta^k \subset \Lambda^k \in [T^k]$, then no node $\subset \lambda(\Lambda^k)$ has infinitely many derivatives along Λ^k lying in the δ^k-block.

DEFINITION 2.7.8 (Standard Derived Assignment). Suppose that $k < n$, and fix $\delta^k \subseteq \eta^k \in T^k$ such that δ^k begins a block and η^k lies in the δ^k-block.

Let η^{k+1} be the shortest η^k-consistent node of T^{k+1} that has no derivative in the δ^k-block. Then the *standard derived assignment* defines $\mathrm{up}(\eta^k) = \eta^{k+1}$. η^k is said to be an *α-node* if $\mathrm{up}(\eta^k)$ is an α-node.

It is easily verified by induction on $n-k$ that the standard derived assignment satisfies clauses (i) and (iii) of Definition 2.7.7; clause (ii) will follow from Lemma 2.12.2 (Derived Assignment).

We note that the assignment of requirements to trees can also be carried out using a simultaneous induction on all trees. This will be of use to us in finding certain parameters when we decompose directing sentences and action sentences. To see this claim, we note that if an algorithm is specified for an assignment, then the choice of requirement for $\eta^k \in T^k$ depends only on the requirements assigned to nodes $\subseteq \lambda(\eta^k)$ and the requirements assigned to nodes $\subset \eta^k$. Once requirements have been assigned to all $\xi^{k+1} \in T^{k+1}$ such that $\xi^{k+1} \subseteq \lambda(\eta^k)$, the algorithm determines the requirement that is assigned to η^k.

2.8. The Weight Function

When $\eta^k \in T^k$ is specified, $\mathrm{out}(\eta^k)$ will provide the information about the method of satisfaction of each requirement assigned to some $\xi^k \subset \eta^k$. However, as it will generally not be the case that $\mathrm{up}(\mathrm{out}(\eta^k)) = \eta^k$ (this is because $\mathrm{out}(\eta^k)$ need not be a derivative of η^k), $\mathrm{out}(\eta^k)$ will not provide information about where we deal with the requirement R_{η^k} assigned to η^k; rather this will occur at initial derivatives of η^k that will usually be located at proper extensions of $\mathrm{out}(\eta^k)$. This process will continue until we reach an initial derivative of η^k on T^0. When $k \leq 2$, η^k will have at most one initial derivative along any $\Lambda^0 \in [T^0]$. However, the uniqueness vanishes when $k \geq 3$, as is illustrated in the following example.

EXAMPLE 2.8.1. We present sets of nodes, closed under initial segments and lying on the trees T^j for $j \leq 3$ in such a way that the node $\rho^3 \in T^3$ has two initial derivatives along the path $\Lambda^0 \in [T^0]$. We suggest that the reader draw the trees in order to be more easily able to follow the description.

We begin with the initial nodes η_0^j of T^j for $j \leq 3$, and specify that for all $j \leq 2$, $\mathrm{up}^{j+1}(\eta_0^j) = \eta_0^{j+1}$. (As is our custom, when a superscript is omitted, it is to be assumed that the superscript should be 0.) $\eta_1 = \eta_0{}^\frown\langle 0\rangle$ will be the immediate successor of η^0 along Λ^0, and so η_0^j will have Σ outcome along $\lambda^j(\eta_1)$ for j even, and will have Π outcome along $\lambda^j(\eta_1)$ for j odd. We will set $\eta_1^j = \lambda^j(\eta_1)$ for $j \leq 2$, $\rho^3 = \lambda^3(\eta_1)$, and specify that $\eta_1^1 = \mathrm{up}(\eta_1)$ and $\eta_1^2 = \mathrm{up}(\eta_1^1)$.

The immediate successor of η_1 along Λ^0 will be $\rho_2 = \eta_1{}^\frown\langle 0\rangle$; thus η_1^j will have Σ outcome along $\lambda^j(\rho_2)$ for $j = 0, 2$, Π outcome for $j = 1$, and $\lambda^3(\rho_2) = \rho^3$. We will set $\rho^j = \lambda^j(\rho_2)$ for $j = 1, 2$, and determine that $\mathrm{up}(\rho_2) = \rho^1$, and

$\text{up}(\rho^j) = \rho^{j+1}$ for $j = 1, 2$. The outcome given ρ_2 along its immediate successor $\eta_3 \subset \Lambda^0$ is irrelevant; in both cases, $\lambda^j(\eta_3)$ will be an immediate successor of ρ^j for $j = 1, 2, 3$. Note thatρ_2 is an initial derivative of ρ^3 along Λ^0.

We will specify that $\text{up}(\eta_3) = \eta_1^1$ and that $\eta_4 = \eta_3{}^\frown\langle\infty\rangle$ is the immediate successor of η_3 along Λ^0. Thus η_4 will switch the outcomes of η_1^1, η_1^2 and η_0^3; in particular, η_1^1 will have Σ outcome along $\lambda(\eta_4)$, η_1^2 will have Π outcome along $\lambda^2(\eta_4)$, and η_0^3 will have Σ outcome along $\lambda^3(\eta_4)$. We set $\eta_2^1 = \lambda(\eta_4)$, and specify that $\text{up}(\eta_4) = \eta_2^1$ and that $\text{up}(\eta_2^1) = \eta_1^2$.

$\rho_5 = \eta_4{}^\frown\langle 0\rangle$ will be the immediate successor of η_4 along Λ^0. η_2^1 will be the immediate predecessor of $\rho_3^1 = \lambda(\rho_5)$ along which it will have Π outcome, and we will set $\text{up}(\rho_5) = \rho_3^1$. $\lambda^2(\rho_5) = \rho_2^2$ will be a new immediate successor of η_1^2 along which η_1^2 has Σ outcome; in particular, $\rho_2^2 | \rho^2$. We set $\text{up}(\rho_3^1) = \rho_2^2$ and $\text{up}(\rho_2^2) = \rho^3$, noting that $\lambda^3(\rho_5) = \rho^3$. We now note that ρ_2^2 is the initial derivative of ρ^3 along $\lambda^2(\rho_5)$, ρ_3^1 is the initial derivative of ρ_2^2 along $\lambda(\rho_5)$, and ρ_5 is the initial derivative of ρ_3^1 along ρ_5, so ρ_5 is another initial derivative of ρ^3 along Λ^0, as we wished to show.

The *weight function*, defined below, will supply bounds and parameters for sentences associated with the requirements to be satisfied in order to prove a given theorem. Most of the theorems that we prove use the weight function only on trees T^i for $i \leq 2$. For these proofs, we introduce a weight function of one variable, a variable representing a node of a tree. In Chapter 10, however, we will need to use the function on higher-level trees; we will define a revised weight function in Section 9.5 as a function of two variables, a node of a tree and one of its initial derivatives.

We now define a one-to-one partial *weight* function, wt, from elements of $\bigsqcup\{T^k : 0 \leq k \leq n\}$ into $\mathbb{N}$. (We take the disjoint union here, differentiating between the empty nodes of the various trees.) This function will serve several purposes that we describe:

- If $\sigma^{k-1} \in T^{k-1}$ is derived from $\sigma^k \in T^k$, then the sentence assigned to σ^{k-1} will frequently be obtained from the sentence φ assigned to σ^k by bounding the leading block of unbounded like quantifiers of φ by $\text{wt}(\sigma^{k-1})$. If $k - 1 = 0$, then the bound $\text{wt}(\sigma^0)$ will be a "stage" of the construction.
- When an argument x for a functional needs to be chosen at $\sigma^k \in T^k$, then we will frequently choose $x = \text{wt}(\sigma^k)$.
- When we are defining a new axiom for a functional at $\sigma^k \in T^k$, then we will frequently specify the use of this axiom as the weight of a node obtained as a function of σ^k.

The properties of the weight function will be used to show that constructions in which action is determined by weight are able to protect axioms. Thus it will never be necessary to "initialize" a construction, as the properties of constructions yielded by initialization will, instead, be consequences of properties of the weight function.

Rather than giving a definition of the weight function, we will merely specify its properties. We leave it to the reader to show that such a function exists. The definition of such a function can be carried out routinely by induction.

DEFINITION 2.8.2. We note that a partial one-to-one weight function wt, with computable domain, can be defined to satisfy the properties listed below. The function maps nodes of $\bigsqcup\{T^k : k \leq n\}$ to the natural numbers. We require that the following properties are satisfied for all $\sigma^k, \tau^k \in T^k$:

$$\sigma^k \subset \tau^k \to \mathrm{wt}(\sigma^k) < \mathrm{wt}(\tau^k). \tag{2.8.1}$$

$$k > 0 \to \mathrm{wt}(\mathrm{out}(\sigma^k)) < \mathrm{wt}(\sigma^k). \tag{2.8.2}$$

$$k > 0 \ \&\ \mathrm{out}(\sigma^k) \subset \mathrm{out}(\tau^k) \to \mathrm{wt}(\sigma^k) < \mathrm{wt}(\tau^k). \tag{2.8.3}$$

We note an important relationship between the functions wt and λ. Suppose that $k < n$, $\sigma^k \subset \tau^k \in T^k$, and $\lambda(\sigma^k) \neq \lambda(\tau^k)$. By Remark 2.4.2(i), $\mathrm{out}(\lambda(\sigma^k)) \subseteq \sigma^k$ and $\mathrm{out}(\lambda(\tau^k)) \subseteq \tau^k$, so by Definition 2.4.1 and as $\sigma^k \subset \tau^k$ and $\lambda(\sigma^k) \neq \lambda(\tau^k)$, $\mathrm{out}(\lambda(\sigma^k)) \subseteq \sigma^k \subset \mathrm{out}(\lambda(\tau^k)) \subseteq \tau^k$. It now follows from (2.8.3) that for all $k < n$:

$$\sigma^k \subset \tau^k \ \&\ \lambda(\sigma^k) \neq \lambda(\tau^k) \to \mathrm{wt}(\lambda(\sigma^k)) < \mathrm{wt}(\lambda(\tau^k)). \tag{2.8.4}$$

2.9. Directing and Action Sentences: An Example

In order to facilitate the use of weights of nodes in sentences, we make the following definition.

DEFINITION 2.9.1. Suppose that a given construction has true path $\Lambda^i \in [T^i]$ for $i \leq n$, and let α either be a node of T^k or a parameter to be interpreted as a node of T^k for some $k > 0$. Define $V_\alpha = \{\eta^{k-1} \subset \Lambda^{k-1} : \mathrm{up}(\eta^{k-1}) = \alpha\}$, and $\widetilde{V}_\alpha = \{\eta^{k-1} \subset \Lambda^{k-1} : \mathrm{up}(\eta^{k-1}) = \alpha \ \&\ \eta^{k-1}$ is Λ^{k-1}-free$\}$.

We will frequently allow quantifiers to range over nodes of V_α. If $\alpha \subset \Lambda^k$, then α will have longer and longer derivatives along Λ^{k-1} until, and if, one of its derivatives has Π outcome. Thus the restriction to these sets will not interfere with the satisfaction of requirements.

We now wish to place conditions on directing sentences and action sentences. Before doing so, some examples are provided to demonstrate how directing sentences and action sentences are defined for a Friedberg–Mučnik requirement. A typical translation might go as follows.

EXAMPLE 2.9.2. The Friedberg–Mučnik sentence $\Phi(A; x) = 0$ will be separated into two parts; the first (the directing sentence) checks to see if a computation occurs at some stage $\mathrm{wt}(\eta)$, and the second (part of an action sentence) preserves this computation, if found. Suppose that the sentence is assigned to $\eta^1 \in T^1$. The first part is

$$\exists u \exists \eta^0 \in V_{\eta^1}(\Phi(A \upharpoonright u; \mathrm{wt}(\eta^1))[\mathrm{wt}(\eta^0)] \downarrow = 0). \tag{2.9.1}$$

The importance of sentences of this type is their openness; once η^0 and u are found satisfying $\Phi(A \upharpoonright u; \mathrm{wt}(\eta^1))[\mathrm{wt}(\eta^0)] \downarrow = 0$, it is possible to force

$$\forall t \geq \mathrm{wt}(\eta^0)(\Phi(A \upharpoonright u; \mathrm{wt}(\eta^1))[\mathrm{wt}(\eta^0)] \downarrow = 0 \,\&\, A^t \upharpoonright u = A^{\mathrm{wt}(\eta^0)} \upharpoonright u)$$

by preventing numbers $\leq u$ from entering A, and so to preserve the truth of the sentence $\Phi(A; \mathrm{wt}(\eta^1)) \downarrow = 0$.

If the sentence $S_{\eta^{k+1}}$ is assigned to $\eta^{k+1} \in T^{k+1}$, then the sentences S_{η^k} that are assigned to derivatives $\eta^k \in T^k$ of η^{k+1} are frequently obtained by bounding some of the quantifiers in $S_{\eta^{k+1}}$. $S_{\eta^{k+1}}$ will be assumed to be in *normal form* (defined below), and we will require the process for bounding quantifiers to be monotonic.

DEFINITION 2.9.3. The *normal form* for $S_{\eta^{k+1}}$ begins with a block of bounded quantifiers, followed by a block of unbounded quantifiers and a final quantifier-free formula. The block of unbounded quantifiers is partitioned into sub-blocks (maximal intervals) of like quantifiers (all universal, or all existential).

EXAMPLE 2.9.4. Consider the Friedberg–Mučnik sentence of (2.9.1) which is in normal form. There is no leading block of bounded quantifiers, so the sentence begins with the block $\exists u \exists \eta^0 \in V_{\eta^1}$ of unbounded quantifiers, which is followed by the quantifier-free formula $\Phi(A \upharpoonright u; \mathrm{wt}(\eta^1))[\mathrm{wt}(\eta^0)] \downarrow = 0$. The block of unbounded quantifiers has one sub-block, the maximal interval $\exists u \exists \eta^0 \in V_{\eta^1}$ of like (existential) quantifiers.

The process of transferring sentences from nodes of a basic module to nodes of a tree is straightforward, and is described in the next definition.

DEFINITION 2.9.5 (Initial Sentence Assignment). When transferring a sentence from a level k node α of a basic module to a node $\eta^k \in T^k$, all occurrences of α are replaced with η^k.

In order to have the proper implications between sentences at successive levels, we will need to require that the sentence decomposition process be monotonic.

DEFINITION 2.9.6 (Monotonic Sentence Decomposition). Let S be a sentence expressed in normal form, and let S_{η^k} be a sentence obtained by bounding each quantifier Q in the leading block of unbounded quantifiers in S by a parameter $p_Q(\eta^k)$ that depends (effectively) on $\eta^k \in T^k$. We say that the decomposition from S to S_{η^k} is *monotonic* if for each such parameter $p = p_Q$, $\eta^k, \rho^k \in T^k$, and $\Lambda^k \in [T^k]$, the following conditions hold:

(i) If $\eta^k \subset \rho^k$, then $p(\eta^k) \leq p(\rho^k)$.
(ii) $\lim\{p(\tau^k) : \tau^k \subset \Lambda^k\} = \infty$.

Instead of bounding an existential quantifier, a witness for the corresponding variable may be chosen; we require that there be an algorithm for the choice of the witness, and the witness chosen to be less than or equal to the bound specified by the parameter for that witness. Furthermore, we allow the quantifier $\exists s$ that ranges over stages to range over the weights of a prescribed computable set of nodes $\subset \Lambda^0$; this set of stages must be infinite if no witness is found.

EXAMPLE 2.9.7. If S is the Friedberg–Mučnik sentence of (2.9.1) and $\eta \in T^0$, then the sentence S_η is obtained from S as follows:

$$\Phi(A \upharpoonright \mathrm{wt}(\lambda(\eta)); \mathrm{wt}(\eta^1))[\mathrm{wt}(\eta)] \downarrow = 0. \tag{2.9.2}$$

Monotonicity follows from (2.8.1) as long as $|\lambda(\Lambda^0)| = \infty$.

Sentences expressing requirements are normally decomposed into three parts, the directing sentence, the activated action sentence, and the validated action sentence. For Friedberg–Mučnik requirements, the corresponding sentences on T^1 and T^0 are now described.

EXAMPLE 2.9.8. The sentence presented in (2.9.1) is the directing sentence for the Friedberg–Mučnik requirement assigned to $\eta^1 \in T^1$. The activated action sentence is $\mathrm{wt}(\eta^1) \notin B$, and the validated action sentence at $\tau^1 \supset \eta^1$ is $\mathrm{wt}(\eta^1) \in B \;\&\; \forall t \geq \mathrm{wt}(\eta^0)(A^t \upharpoonright u = A^{\mathrm{wt}(\eta^0)} \upharpoonright u)$, where η^0 and u are the witnesses obtained from the directing sentence. When these sentences are decomposed for $\eta^0 \in T^0$ at $\rho^0 \supset \eta^0$, the directing sentence is that presented in (2.9.2), the activated action sentence at ρ^0 is $\mathrm{wt}(\eta^1) \notin B^{\mathrm{wt}(\rho^0)}$, and the validated action sentence at $\rho^0 \supset \eta^0$ is $\mathrm{wt}(\eta^1) \in B^{\mathrm{wt}(\rho^0)} \;\&\; \forall t \in [\mathrm{wt}(\eta^0), \mathrm{wt}(\rho^0)](A^t \upharpoonright u = A^{\mathrm{wt}(\eta^0)} \upharpoonright u)$. We will always allow witnesses for existential quantifiers in the directing sentence to be used as parameters for action sentences.

2.10. Directing and Action Sentences: Properties

Three sentences will be associated with each non-terminal node α of a basic module, and α will have a *level*, written as $\mathrm{lev}(\alpha)$, determined by the logical complexity of these sentences. The homomorphism of Definition 2.7.1 transfers this level to nodes of T^n; directing sentences and action sentences are also transferred, possibly with a reinterpretation of parameters. The *level* of a derivative is the same as the level of its antiderivative on T^n. The *level* of a basic module is the maximum of the levels of its nodes, and is also the *level* of the requirement for which the basic module was established.

The first sentence, called the *directing sentence*, will determine the outcome for a node during the construction, and thereby determine the action to be taken for that node. Given a directing sentence S_η for the node $\eta = \eta^k \in T^k$, action will consist of declaring axioms with value depending on the truth or falsity of S_η, or placing numbers into sets. If S_η is false, then an activated outcome for η^k will be followed along successors ρ^k of η^k, and *activated action* for η^k will be implemented at certain ρ^k; and if S_η is true, then a validated outcome for η^k will be followed along successors ρ^k of η^k, and *validated action* for η^k will be implemented at certain ρ^k. If $\Lambda^k \in [T^k]$ and $S \subseteq (\eta^k, \Lambda^k)$, then we say that action for η^k along Λ^k is *supported* at S if S is the subset of Λ^k at which the construction implements action for η^k. (For example, a thickness requirement for η^1 might say that all elements of a computable set R that are $\geq \mathrm{wt}(\eta^1)$ are to be placed into a set A; when this action is supported at $\rho^1 \supset \eta^1$, the action will be to place all $x \in R \cap [\mathrm{wt}(\eta^1), \mathrm{wt}(\rho^1))$ into A.) The action for η^k that is *supported along* Λ^k is the union of the action sentences for η^k that are supported along Λ^k at a node in the support set $S \subseteq \Lambda^k$. The directing sentence for $\eta^0 \in T^0$ and the action sentences for η^0 at $\rho^0 \in T^0$ will specify action that can be carried out effectively. The support set S for $\eta^k \subseteq \Lambda^k$ will be chosen to have sufficiently many nodes so that the union of the action taken for η^k at $\rho^k \in S$ will imply the union of the action taken for $\mathrm{up}(\eta^k)$ at its support set $S^{k+1} \subseteq \Lambda^{k+1} = \lambda(\Lambda^k)$. Action for η at ρ is always implemented before the truth value of the directing sentence for ρ is determined.

Directing sentences and action sentences will normally be decomposed by bounding quantifiers when passing from tree to tree, finally producing quantifier-free sentences for nodes of T^0. We allow witnesses for existential quantifiers in the directing sentence to be used as parameters for action sentences. The decomposition of sentences must satisfy certain conditions that are specified in the next definition. Condition (i) is the *initial condition*, requiring that the truth values of the initial directing sentences and action along the true path ensure the satisfaction of the requirement for which they have been introduced; conditions (ii) and (iii) are the *pulldown conditions* and ensure a faithful passage from the truth value of the directing sentence and action sentences on T^k to the truth value of the directing sentence and action sentences on T^{k+1}; and condition (iv) is the *final condition*, requiring the construction on T^0 to faithfully reflect the truth value of the directing sentences and action on T^0. These conditions are now stated.

DEFINITION 2.10.1 (Directing Sentence and Action Conditions). The following restrictions are imposed on directing sentences and action sentences:

(i) If $\mathrm{lev}(\eta^k) = k$ and η^k is Λ^k-free and is activated (validated, resp.) along Λ^k, then the action for η^k that is supported along Λ^k and the falsity (truth, resp.) of the directing sentence S_{η^k} for η^k ensure the satisfaction of the requirement R assigned to η^k.

(ii) Fix $k < n$ and $\Lambda^k \in [T^k]$, and let $\eta^{k+1} \subset \Lambda^{k+1} = \lambda(\Lambda^k)$ be given such that $\text{lev}(\eta^{k+1}) \geq k+1$. Suppose that η^{k+1} is Λ^{k+1}-free. Let $S_{\eta^{k+1}}$ be the directing sentence for η^{k+1}, and for each derivative η^k of η^{k+1} along Λ^k, let S_{η^k} be its directing sentence. Then if η^{k+1} is validated (activated, resp.) along Λ^{k+1}, then the (possibly infinite) conjunction of the sentences S_{η^k} ($\neg S_{\eta^k}$, resp.) as η^k ranges over the Λ^k-free derivatives of η^{k+1} must imply $S_{\eta^{k+1}}$ ($\neg S_{\eta^{k+1}}$, resp.).

(iii) Fix $k < n$ and $\Lambda^k \in [T^k]$, and let $\eta^{k+1} \subset \Lambda^{k+1} = \lambda(\Lambda^k)$ be given such that $\text{lev}(\eta^{k+1}) \geq k+1$. Suppose that η^{k+1} is Λ^{k+1}-free. Then if η^{k+1} is validated (activated, resp.) along Λ^{k+1}, then the (possibly infinite) conjunction of the action sentences for η^k supported along Λ^k as η^k ranges over the Λ^k-free derivatives of η^{k+1} must imply the action sentences for η^{k+1} that are supported along Λ^{k+1}.

(iv) Let $\Lambda^0 \in [T^0]$ be the true path for a construction, and let $\eta \subset \rho \subset \Lambda^0$ be given. Let S_η be the directing sentence for η, and let $A_{\eta,\rho}$ be the action sentence for η. Assume that action for η is supported at ρ. Then if η is Λ^0-free, then:
  (a) η is validated along Λ^0 iff S_η is true.
  (b) $A_{\eta,\rho}$ is true.

2.11. The Framework Theorem

The constructions that we present beginning in the next chapter will all satisfy certain basic properties, but we will have to verify that this is the case. Part of the verification will be by inspection, but another part of the verification will rely on lemmas proved for constructions which are assumed to have some of these properties. Thus we will have to take care to avoid circular arguments. We now indicate how this is done.

The lemmas that are proved assume that all constructions satisfy the properties introduced in Sections 2.3–2.8. In particular, the conditions of Definition 2.3.1 (Basic Modules), Definition 2.7.2 (Initial Assignment), and Definition 2.7.7 (Derived Assignment) will be used in the proofs of the lemmas of Chapters 8 and 9, and when we use the term *effective construction* or just *construction*, we will be referring only to constructions satisfying these properties. Within constructions, the verification that the basic module conditions (Definition 2.3.1(i)–(iii)) are satisfied will be by inspection, the verification that the initial assignment condition (Definition 2.7.2(i)) is satisfied will proceed through an analysis of the requirements and will not use any of the lemmas, and the verification that the derived assignment conditions (Definition 2.7.7(i)–(iii)) are satisfied will be an appeal to one of the derived assignment lemmas, namely to Lemmas 2.12.2 (Standard Derived Assignment), Lemma

8.6.2 (Preferential Derived Assignment), or Lemma 8.6.5 (Module-Respecting Derived Assignment). Furthermore, in the proofs of these lemmas, clauses (i) and (iii) of Definition 2.7.7 are verified by inspection, and the proof of clause (ii) of Definition 2.7.7 does not involve the use of a lemma whose proof uses this condition. (Note that the proof of Lemma 2.12.2 uses Lemma 2.12.1(i), but the proof of the latter statement uses only clause (i) of Definition 2.7.7, which is allowable.)

We now turn to the Framework Theorem, which stipulates that constructions succeed in satisfying all requirements if they satisfy certain basic properties, properties that are common to all priority method constructions. We then prove that the satisfaction of some of these properties is immediate if the sentence decomposition process is monotonic.

Assume that Λ^0 is the true path for a construction, and for all $j \leq n$, set $\Lambda^j = \lambda^j(\Lambda^0)$.

THEOREM 2.11.1 (Framework Theorem). *Suppose that an effective construction satisfying clauses (i)–(iv) of Definition* 2.10.1 *is given. Then all requirements are satisfied.*

PROOF. By clause (i) of Definition 2.10.1, it suffices to show that for all Λ^n-free nodes η^n, if $\text{lev}(\eta^n) = k$ and η^k is the unique (by (2.6.3)) derivative of η^n along Λ^k, then if the directing sentence S_{η^k} is true, then validated action for η^k is carried out along Λ^k, and if S_{η^k} is false, then activated action for η^k is carried out along Λ^k. Fix a Λ^n-free node η^n. By repeated induction on $i = n - j$, it follows from clause (ii) of Definition 2.10.1 that if S_{η^k} is true (false, resp.), then the directing sentences S_{η^0} are true (false, resp.) for all Λ^0-free derivatives η^0 of η^k. By clause (iv) of Definition 2.10.1, for all Λ^0-free derivatives η^0 of η^k, S_{η^0} is true iff η^0 is validated along Λ^0. By repeated induction on $i = n - j$, using clauses (iii) and (iv) of Definition 2.10.1, it follows that validated (activated, resp.) action for η^k along Λ^k will be implemented if validated (activated, resp.) action is implemented for all Λ^0-free derivatives η^0 of η^k, which in turn will happen if η^0 is validated (activated, resp.) along Λ^0. ⊣

Clauses (i)–(iv) of Definition 2.10.1 permit the analysis of a construction to concentrate on Λ^0-free nodes, and provide a common framework for priority arguments. Furthermore, clauses (ii) and (iii) of Definition 2.10.1 will hold whenever clauses (i)–(iii) of Definition 2.3.1 and clauses (i)–(iii) of Definition 2.7.7 hold and the sentence decomposition is monotonic, the construction has infinite support, and support propagates, as is stated in the Monotonic Sentence Decomposition Lemma below (proved as Lemma 8.6.1).

The following definitions introduce properties needed in the hypothesis of the Monotonic Sentence Decomposition Lemma.

DEFINITION 2.11.2. We say that a construction has *infinite support* if for all $k < n$, whenever $\Lambda^k \in [T^k]$, $|\Lambda^k| = \infty$, and $\eta^k \subset \Lambda^k$ is Λ^k-free, then the support set for η^k along Λ^k is infinite.

DEFINITION 2.11.3. We say that support for a construction *propagates* if whenever $k < n$, $\eta^k \subset \rho^k \in T^k \cup [T^k]$ is ρ^k-free and action for up(η^k) is supported at $\lambda(\rho^k)$, then action for η^k is supported at ρ^k.

LEMMA 2.11.4 (Monotonic Sentence Decomposition Lemma). *Suppose that a construction satisfies clauses (i)–(iii) of Definition* 2.3.1 *and clauses (i)–(iii) of Definition* 2.7.7, *and that* $\Lambda^k \in [T^k]$ *and* $\eta^{k+1} \in T^{k+1}$ *are given such that* η^{k+1} *is* $\lambda(\Lambda^k)$*-free. Then clause (ii) of Definition* 2.10.1 *is satisfied whenever a monotonic sentence decomposition process for directing sentences is followed, and clause (iii) of Definition* 2.10.1 *is satisfied whenever a monotonic sentence decomposition process for action sentences is followed, the construction has infinite support, and support propagates.* ⊣

The Framework Theorem reduces the proof of the success of a priority method construction to the verification of clauses (i)–(iv) of Definition 2.10.1. Definition 2.10.1(i) can be verified by inspection and an analysis of the basic modules. By Lemma 2.11.4 (Monotonic Sentence Decomposition), Definition 2.10.1(ii),(iii) will follow if Definition 2.3.1(i)–(iii) and clauses (i)–(iii) of Definition 2.7.7 hold, sentence decomposition is monotonic, the construction has infinite support and support propagates.

Clauses (i)–(iii) of Definition 2.3.1 impose restrictions on the form of the basic module, and follow by inspection. By Lemma 2.12.2 (Derived Assignment), Definition 2.7.7(i)–(iii) follow from the implementation of the standard derived assignment, so we will follow this assignment whenever possible. The monotonic sentence decomposition and the infinite support and propagation of support conditions will be easily checked, sometimes by applying lemmas about the framework. However, the precise monotonic decomposition employed and the support set must be carefully chosen to be compatible with the properties needed to verify Definition 2.10.1(iv).

The above discussion indicates that all conditions, with the possible exception of Definition 2.10.1(iv), needed to verify the hypothesis of the Framework Theorem can quickly be verified by inspection under a suitable implementation of the framework, and determine a construction that is to be implemented on T^0. Clause (iv) of Definition 2.10.1 requires the construction on T^0 to be *implementable*, i.e., that there be no conflicts preventing nodes of T^0 from implementing the prescribed action. We will always require that the construction choose its true path in accordance with Definition 2.10.1(iva). (When a construction implements backtracking, the outcome of certain nodes is forced independently of the truth of the directing sentence, but no node whose outcome is forced in this way will be Λ^0-free.) The verification of Definition 2.10.1(iv) for a given construction can be separated into two parts. The first part requires a characterization of the nodes of T^0 at which conflicts to the implementation of action might possibly occur. Sentence assignment allows us to easily identify these nodes. The second part of the verification of Definition 2.10.1(iv) involves showing that no conflicts occur. This part is carried out through a comparison of parameters specified for the monotonic decomposition and weights for the nodes that produce the potential conflict (as determined in the first part) and will frequently require some of the systems of trees machinery. Thus the key ingredients for the success of the construction are the design of the basic module, and the choice of parameters for the monotonic sentence decomposition that allow Definition 2.10.1(iv) to be verified.

We conclude that we can apply the following modified framework theorem when additional conditions are met.

THEOREM 2.11.5 (Modified Framework Theorem). *Fix an effective construction satisfying clauses (i)–(iii) of Definition* 2.3.1 *and clauses (i) and (iv) of Definition* 2.10.1, *and suppose that clauses (i)–(iii) of Definition* 2.7.7 *hold (which will be the case if the standard derived assignment is followed), that sentence decomposition is monotonic, that the construction has infinite support, and that support propagates. Then all requirements are satisfied.*

2.12. A Sampling of Lemmas

As we mentioned earlier, we plan to separate the lemmas delineating properties of the framework from the proofs of theorems about degrees using these lemmas. Thus in Chapters 3 through 7, we will just state the lemmas about the framework needed for those sections, and present the proofs of the theorems about degrees using those lemmas; the lemmas will be proved in Chapters 8 and 9. Some readers, however, may want to develop a feeling for the operation of the framework before beginning Chapter 3; to accommodate these readers, we will present proofs of two lemmas about the framework in this section, the Limit Path and Nesting Lemmas, as well as a proof of the Derived Assignment Lemma. Recall that we are allowing the use of all conditions required of constructions mentioned in Sections 2.2–2.8 in the proofs of the Limit Path and Nesting Lemmas.

The path generation process is designed to have many nice, uniform properties. We describe some of those properties in the next lemma. This lemma states that paths through the trees are either infinite or end at nodes of the tree to which no requirements are assigned, and that initial and principal derivatives exist. This lemma is used frequently to prove other lemmas about the framework, and provides an important tool in the analysis of derived assignments.

LEMMA 2.12.1 (Limit Path Lemma). *Fix $k \in [0, n)$ and $\rho^k \in T^k \cup [T^k]$. Let $\rho^{k+1} = \lambda(\rho^k)$ if $|\rho^k| < \infty$, and let $\rho^{k+1} = \lim\{\lambda(\eta^k) : \eta^k \subset \rho^k\}$ if $|\rho^k| = \infty$. Then:*

(i) *If $\eta^{k+1} \subset \lambda(\rho^k)$, then η^{k+1} has an initial derivative η^k along ρ^k and $\lambda(\eta^k) = \eta^{k+1}$.*

(ii) *If $\eta^{k+1} \subset \lambda(\rho^k)$, then there is a $\nu^k \subseteq \rho^k$ such that $\pi^k = (\nu^k)^-$ is the principal derivative of η^{k+1} along ρ^k, $(\lambda(\nu^k))^- = \eta^{k+1}$, and for all $\xi^k \subseteq \rho^k$, $\lambda(\nu^k) \subseteq \lambda(\xi^k)$ iff $\nu^k \subseteq \xi^k$.*

(iii) *If $|\rho^k| = \infty$, then either $|\rho^{k+1}| = \infty$, or no $\xi^{k+1} \supseteq \rho^{k+1}$ has an antiderivative $\xi^k \subset \rho^k$.*

(iv) *If $|\rho^k| = \infty$, then every block begun at $\delta^k \subset \rho^k$ is completed along ρ^k.*

PROOF. We prove the lemma for $k = 0$, noting that the proof relativizes. As is our convention, we will drop the superscript 0.

(i): By Remark 2.4.2(iii) and as $\eta^1 \subset \lambda(\rho)$, η^1 must have a derivative along ρ; the shortest derivative η of η^1 along ρ is the initial derivative of η^1 along ρ. Now (2.6.1) tells us that $\lambda(\eta) \supseteq \eta^1$, and by Remark 2.4.2(iii), $\lambda(\eta) \not\supset \eta^1$. Hence $\lambda(\eta) = \eta^1$.

(ii): Suppose that $\eta^1 \subset \lambda(\rho)$. By Remark 2.4.2(iii), η^1 must have a derivative $\subset \rho^1$, hence must have a principal derivative $\pi \subset \rho$. Fix $\nu \subseteq \rho$ such that $\nu^- = \pi$. By (2.6.1), $\lambda(\pi) \supseteq \eta^1$, and so by Definition 2.4.1, $\lambda(\nu) \supseteq \eta^1{}^\frown\langle\nu\rangle$. Let $\nu^1 = \eta^1{}^\frown\langle\nu\rangle$. By (2.6.1) and Definition 2.4.1, if $\mathrm{up}(\xi) = \eta^1{}^\frown\langle\nu\rangle$ then $\nu \subseteq \xi$. Hence again by Definition 2.4.1, we cannot have $\lambda(\xi) \supset \nu^1$ for any $\xi \subsetneq \nu$. We thus conclude that $\lambda(\nu) = \nu^1$.

As π is the principal derivative of η^1 along ρ and $\eta^1 \subset \lambda(\rho)$, we must have $\nu^1 \subseteq \lambda(\rho)$; and we have already noted that $\nu^1 = \lambda(\nu)$. It now follows from Remark 2.4.2(ii) that $\nu^1 \subseteq \lambda(\xi)$ for all $\xi \in [\nu, \rho]$. By Definition 2.4.1, if $\xi \subset \nu$ then $\lambda(\xi) \not\supseteq \nu^1$, so (ii) follows.

(iii): Suppose that $|\rho| = \infty$ and $|\rho^1| < \infty$. By (2.6.1) and Definition 2.4.1, there can be no $\sigma \subset \rho$ such that $\mathrm{up}(\sigma) \supset \rho^1$ unless there is a $\xi \subset \rho$ such that $\mathrm{up}(\xi) = \rho^1$. Hence it suffices to show that there is no $\xi \subset \rho$ such that $\mathrm{up}(\xi) = \rho^1$. We assume that such a ξ exists, and obtain a contradiction.

Fix the shortest $\eta \subset \rho$ such that $\lambda(\eta) \supseteq \rho^1$. By Remark 2.4.2(ii), $\lambda(\tau) \supseteq \rho^1$ for all $\tau \in [\eta, \rho)$. As there is a $\xi \subset \rho$ such that $\mathrm{up}(\xi) = \rho^1$, ρ^1 must have a principal derivative $\pi \subset \rho$; fix $\nu \subset \rho$ such that $\nu^- = \pi$. By (ii), $\lambda(\tau) \supseteq \rho^1{}^\frown\langle\nu\rangle$ for all $\tau \in [\nu, \rho)$, contrary to our definition of $\rho^1 = \lim\{\lambda(\eta) : \eta \subset \rho\}$, so (iii) follows.

(iv): We proceed by reverse induction on $j < n$, so may assume that (iv) holds for $j = 1$ and verify (iv) for $j = 0$. (Note that if $n = 1$, then either ρ^1 is finite, or (iv) follows from clause (i) of Definition 2.3.1 and Definition 2.7.1.) Suppose that $|\rho| = \infty$ and a block is begun at $\delta \subset \rho$. By Definition 2.7.5 and (i), blocks on T^0 are completed at nodes that are initial derivatives of nodes of T^1 completing blocks, so this is only possible if $|\lambda(\rho)| < \infty$. Thus by Remark

2.4.2(ii), there is an $\eta \subset \rho$ such that for all ξ satisfying $\eta \subseteq \xi \subseteq \rho$, $\lambda(\xi) = \rho^1$. By clause (ii) of Definition 2.7.7, there is a $\tau \subset \rho$ that is a derivative of the longest node $\subset \rho^1$ having level 1 and Π outcome along ρ^1. By Definition 2.7.5, the δ-block along ρ is completed at such a τ. ⊣

We now turn to a proof of the Derived Assignment Lemma, showing that the standard derived assignment satisfies clauses (i)–(iii) of Definition 2.7.7. Recall that in order to avoid circularity, the proof should avoid the use of any lemma whose proof uses these clauses. However, as clauses (i) and (iii) of Definition 2.7.7 follow easily from the definition of the standard derived assignment without an appeal to any lemma, we are permitted to use Lemmas whose proofs use these clauses. In particular, we may use Lemma 2.12.1(i) (Limit Path) to prove that clause (ii) of Definition 2.7.7 is satisfied, as its proof uses only clause (i) of Definition 2.7.7.

LEMMA 2.12.2 (Derived Assignment Lemma). *The standard derived assignment satisfies clauses* (*i*)–(*iii*) *of Definition* 2.7.7.

PROOF. Clauses (i) and (iii) of Definition 2.7.7 follow easily by induction on $n - k$ from the definition of the standard derived assignment. Suppose that we have a level n construction. We verify clause (ii) of Definition 2.7.7 by induction on $n - k$. Fix $\delta^k \subseteq \eta^k \subset \rho^k \in T^k$ such that $(\rho^k)^- = \eta^k$, and assume that a ρ^k-path through the δ^k-block is completed at η^k. Then $\text{up}^j(\eta^k)$ is an initial derivative for all $j \in [k, n]$, so by Lemma 2.12.1(i) (Limit Path), for all $j \in [k, n]$, $\lambda^j(\eta^k) = (\lambda^j(\rho^k))^-$. Furthermore, $[\mu^j, \pi^j]$ is a $\lambda^j(\rho^k)$-link iff it is a $\lambda^j(\eta^k)$-link. Hence the nodes $\subseteq \lambda(\eta^k)$ that are ρ^k-consistent are also η^k-consistent. Clause (ii) of Definition 2.7.7 now follows from the definition of the standard derived assignment. ⊣

Another key property of the framework is that links along a given path through a tree, or along a given node on a tree, are nested. The following lemma shows this to be the case.

LEMMA 2.12.3 (Nesting Lemma). *Fix $k \leq n$ and $\rho^k \in T^k \cup [T^k]$. Suppose that, for $i \leq 1$, $[\mu_i^k, \pi_i^k]$ is a ρ^k-link and that $\pi_0^k \subset \pi_1^k$. Then either $\mu_1^k \subseteq \mu_0^k$ or $\pi_0^k \subset \mu_1^k$.*

PROOF. We proceed first by induction on $n - k$, then by induction on $|\rho^k|$, and finally by induction on $|\rho^k| - |\pi_0^k|$. As there are no links on T^n, the lemma holds trivially for $k = n$. Assume that the lemma is true for $k + 1$ in place of k. As the proof relativizes, we just present the proof for $k = 0$, and drop the superscript when it is 0.

For each $i \leq 1$, fix the $\lambda(\rho)$-link $[\mu_i^1, \pi_i^1]$ from which $[\mu_i, \pi_i]$ is derived if such a link exists; otherwise, $[\mu_i, \pi_i]$ is a primary ρ-link, and we set $\mu_i^1 = \pi_i^1 = \text{up}(\mu_i) = \text{up}(\pi_i)$. We note that by Definitions 2.4.4 and 2.5.4, for $i \leq 1$, μ_i (π_i, resp.) is the initial (principal, resp.) derivative of μ_i^1(π_i^1, resp.) along ρ. First assume that π_0^1 and π_1^1 are comparable. As $\pi_0 \subset \pi_1$, it follows from Lemma 2.12.1(ii) (Limit Path) that $\pi_0^1 \subset \pi_1^1$. If $\pi_0^1 \subset \mu_1^1$, then by Lemma 2.12.1(i),(ii) (Limit Path), $\mu_0 \subset \pi_0 \subset \mu_1 \subset \pi_1$ and the lemma holds. Otherwise, $\mu_1^1 \subseteq \pi_0^1 \subset \pi_1^1$. If $\mu_0^1 = \pi_0^1$, then $\mu_1^1 \subseteq \mu_0^1$. And if $\mu_0^1 \neq \pi_0^1$, then by induction, $\mu_1^1 \subseteq \mu_0^1$. In either case, it follows from Lemma 2.12.1(i) (Limit Path) that $\mu_1 \subseteq \mu_0$.

Now assume that $\pi_0^1 \mid \pi_1^1$, and let $\kappa^1 = \pi_0^1 \wedge \pi_1^1$. By (2.6.1), $\lambda(\pi_i) \supseteq \pi_i^1$ for $i \leq 1$, so by Remark 2.4.2(vi) and as $\pi_0 \subset \pi_1$, κ^1 has a principal derivative μ that has Σ outcome along π_0 and so is the initial derivative of κ^1 along π_0, and a principal derivative $\pi \in (\pi_0, \pi_1)$ along π_1 that has Π outcome along π_1. Thus $[\mu, \pi]$ is a primary π_1-link that restrains π_0, and so by induction on $|\rho|$, $\mu \subseteq \mu_0$. As $[\mu, \pi]$ is a primary π_1-link and $\pi_1 \subset \rho$, $[\mu, \pi]$ is also a primary ρ-link. Hence by induction on $|\rho| - |\pi_0|$, either $\pi \subset \mu_1$ or $\mu_1 \subseteq \mu$. As $\mu \subseteq \mu_0$ and $\pi_0 \subset \pi$, it follows that either $\pi_0 \subset \mu_1$ or $\mu_1 \subseteq \mu_0$. ⊣

Chapter 3

Σ_1 CONSTRUCTIONS

Several well-known theorems whose proofs rely on priority method constructions at the Σ_1 level are proved in this chapter. It is our intent to separate proofs of priority lemmas from proofs of the satisfaction of requirements. Thus we begin the study of theorems at each level by stating, without proof, a lemma enumerating the properties of the framework needed to implement priority constructions at that level; these lemmas are proved in Chapter 8.

We discuss the nature of Σ_1 constructions and the properties needed to implement them in Section 3.1. These are applied to prove the Friedberg–Mučnik Theorem in Section 3.2, the existence of a low non-zero computably enumerable degree in Section 3.3, and the existence of a properly d-c.e. degree in Section 3.4.

3.1. Σ_1 Constructions

Σ_1 constructions are those that take place on two trees, T^0 and T^1. We begin with a general overview of the types of requirements encountered in proving theorems about $\mathcal{R}$, followed by a discussion of strategies used to satisfy the particular types of requirements appearing in Σ_1 constructions.

Most requirements can be rewritten as sentences built from atomic formulas of the form

$$\Phi(A, V; x) = \Delta(B, W; y) \tag{3.1.1}$$

where A and B are sets being constructed, and V and W are given sets. (A set A is identified with its characteristic function χ_A, allowing the reduction of atomic formulas of the form $x \in A$ to $\chi_A(A; x) = 1$.) Each set or functional occurring in (3.1.1) may be one that is being constructed, or one over which the construction has no control. Thus a construction can be viewed as a two-player game (this approach was introduced by Lachlan [8]), with Player I having responsibility for the sets and functionals that are not controlled by the construction, and Player II controlling the sets and functionals under construction. A typical move for a Σ_1 construction in such a game begins with the discovery of an axiom of the form

$$\Phi(A \upharpoonright u; x)[s] = m \tag{3.1.2}$$

which requires action to declare an axiom of the form

$$\Delta(B \upharpoonright v; y)[t] = k. \tag{3.1.3}$$

(For example, a Friedberg–Mučnik requirement begins with the discovery of the axiom $\Phi(A \upharpoonright u; x)[s] = 0$, and action corresponds to declaring the axiom $\Delta(B \upharpoonright x+1; x)[s+1] = 0$, or the axiom $\Delta(B \upharpoonright x+1; x)[s+1] = 1$ and later, axioms implying $A^t \upharpoonright u = A^s \upharpoonright u$ for infinitely many $t > s$.) Thus the node of the tree of strategies T^1 that has the ultimate responsibility to declare the axiom must ensure that an axiom $\Delta(B \upharpoonright w; y)[p] = r \neq k$ has not previously been declared by another node with $B^p \upharpoonright w$ and $B^t \upharpoonright v$ compatible, i.e., it must be shown that control of the declaration of an axiom resides with the node that is ultimately assigned the responsibility for the axiom. Once an axiom is discovered or declared, the correctness of the axiom must be preserved. Preservation of the axiom $\Phi(A \upharpoonright u; x)[s] = m$ requires *restraint* on A, preventing numbers $< u$ from entering A after the axiom has been declared. And correctness of axioms is ensured by requiring that if the value required for the axiom $\Delta(B; y)$ with use $v - 1$ is *revised* (as might be the case if the axiom is to correctly compute $C(y)$ and y enters C), then some number $< v$ enters B allowing a new axiom to be declared. More intricate strategies will be necessary for requirements placed on higher-level trees.

All sets mentioned in Σ_1 constructions will be controlled by Player II (so are being constructed), but functionals may be controlled by either player. Thus both restraint and revision (henceforth called *correction*) are controlled by Player II. In most constructions that we present, numbers entering sets will be of the form $\mathrm{wt}(\eta^1)$ for some $\eta^1 \in T^1$, and will be greater than the restraint imposed to preserve axioms for $\xi^1 \subset \eta^1$. If an axiom is newly declared in accordance with the action required for η^1 at ρ^1, and if $\nu^1 \subseteq \rho^1$ and $(\nu^1)^- = \eta^1$, then the use of the axiom will be $< \mathrm{wt}(\nu^1)$. A weight computation will show that $\mathrm{wt}(\nu^1) > \mathrm{wt}(\eta^1)$. Thus the failure of restraint resulting in an injury to an axiom can occur only if the current path is switched at some $\kappa^1 \subseteq \eta^1$. If the axiom in question is still required after a switch of paths through T^1, we will try to place $\mathrm{wt}(\kappa^1)$ into the oracle for the axiom, allowing the axiom to be corrected. If this is not possible, then we will show that the axiom originally declared agrees with the axiom required after the current path switches.

While the success of these strategies can be verified directly for each construction, there is a great deal of duplication in the verification of clause (iv) of Definition 2.10.1. We can eliminate much of this duplication by reducing the verification of clause (iv) of Definition 2.10.1 to a condition on the interaction of requirements along paths through T^1, together with a condition characterizing the effect of switching the current path. These can usually be routinely verified through a weight inequality computation, thus further simplifying proofs. The statement of these conditions requires several definitions.

The first definition specifies the atomic formulas of a language in which the sentences governing action for requirements are to be expressed. This language is rich enough to cover statements about $\mathcal{R}$. We will later allow quantifiers to range over sets arising from tree constructions.

DEFINITION 3.1.1. A *condition* is a collection of boolean combinations of formulas of one of the following forms: $x \in A^s$, $x \notin A^s$, or $\Phi^s(A^t \upharpoonright u; x) = m$, where Φ is a computable partial functional, $A \subset \mathbb{N}$, $u, x, s, t \in \mathbb{N}$, and $m \in \mathbb{N} \cup \{\uparrow, \downarrow\}$. The first two forms are called *set conditions* or *set axioms* and the last form is a *functional condition* or *functional axiom*. If a quantifier ranges over a finite set S and for each $s \in S$, the scope of the quantifier is a boolean combination of conditions, then we view each such instance as a condition. A *restraint* condition is one of the form $x \notin A^s$. A set of conditions is *consistent* if the following hold:

(i) It does not simultaneously contain the set conditions $x \in A^s$ and $x \notin A^t$ for some $t \geq s$.
(ii) It does not contain conditions $\Phi^{s_i}(A^{t_i} \upharpoonright u_i; x) = m_i$ for $i \leq 1$ such that $m_0 \neq m_1$ and $A^{t_0} \upharpoonright u_0$ and $A^{t_1} \upharpoonright u_1$ are compatible.

(We assume, in (i) and (ii), that both occurrences of the symbol A represent the same set.)

The constructions that we present will choose a path through T^0 determined by the truth value of directing sentences for nodes encountered along Λ^0. Rather than specifying each construction, we will just require that the construction be *faithfully executed.*

DEFINITION 3.1.2. We say that a construction, whose true path is $\Lambda^0 \in [T^0]$, is *faithfully executed* on T^0 if the following conditions are satisfied for all $\eta \subset \rho \subset \Lambda^0$ such that η is Λ^0-free and action for η is supported at ρ:

(i) If S_η is true, then $\eta^\frown\langle\infty\rangle \subset \Lambda^0$ and validated action for η at ρ is implemented.
(ii) If S_η is false, then $\eta^\frown\langle 0\rangle \subset \Lambda^0$ and activated action for η at ρ is implemented.

Nodes $\rho \subset \Lambda^0 \in [T^0]$ will support action for nodes $\eta \subset \rho$. For each η, we must require that the action instructions supported at ρ are under the control of Player II.

DEFINITION 3.1.3. We say that a construction *admits consistent action along* Λ^0 if for every $\eta \subseteq \Lambda^0 \in [T^0]$, action specified for η can be implemented by Player II, i.e., the following conditions hold for all ρ such that $\eta \subset \rho \subseteq \Lambda^0$:

(i) If action for η supported at ρ specifies that $x \notin A^{\mathrm{wt}(\rho)}$, then there do not exist $\xi \subset \tau \subseteq \rho$ such that action for ξ supported at τ specifies that $x \in A^{\mathrm{wt}(\tau)}$;

(ii) If action for η supported at ρ specifies that $\Phi(A \upharpoonright u; x)[\mathrm{wt}(\delta)] \uparrow$ for some $\delta \subseteq \rho$, then there do not exist $\xi \subset \tau \subseteq \rho$, $\gamma \subseteq \delta$ and v such that $A^{\mathrm{wt}(\delta)} \upharpoonright u$ and $A^{\mathrm{wt}(\gamma)} \upharpoonright v$ are compatible and action for ξ supported at τ specifies that $\Phi(A \upharpoonright v; x)[\mathrm{wt}(\gamma)] \downarrow$;

(iii) If action for η supported at ρ specifies that $\Phi(A \upharpoonright u; x)[\mathrm{wt}(\delta)] \downarrow = m$ for some $\delta \subseteq \rho$, then there do not exist $\xi \subset \tau \subseteq \rho$, $\gamma \subseteq \delta$ and v such that $A^{\mathrm{wt}(\delta)} \upharpoonright u$ and $A^{\mathrm{wt}(\gamma)} \upharpoonright v$ are compatible and action for ξ supported at τ specifies that $\Phi(A \upharpoonright v; x)[\mathrm{wt}(\gamma)] \downarrow = p \neq m$.

Action will be specified at nodes $\rho \subset \Lambda^0$, and will describe conditions that are to be satisfied for a given stage s. We will need to show that we can consistently satisfy these conditions. The next definition picks out pairs of situations that might potentially be inconsistent as they specify action conditions about the membership of a fixed number in a given set or about the value of a functional, albeit at different stages. (Such a situation occurs in the construction of a low computably enumerable set, where all nodes at a given level of T^1 have the responsibility to define an axiom with the same oracle and argument.) These pairs are said to *share action*. Definition 3.1.5 then specifies when shared action can be consistently satisfied.

DEFINITION 3.1.4. Let $\eta_0 \subset \rho_0 \in T^0$ and $\eta_1 \subset \rho_1 \in T^0$ be given. We say that $\langle \eta_0, \rho_0 \rangle$ and $\langle \eta_1, \rho_1 \rangle$ *share action* if either:

(i) There are $x \in \mathbb{N}$ and $A \subset \mathbb{N}$ such that, for $i \leq 1$, action for η_i at ρ_i contains a condition of the form $x \in A^s$ or $x \notin A^t$. (Note that we do not require both action conditions to specify the same stage, nor to specify the same type (positive or negative) of action; but their choice of x and A must agree.)

(ii) There are $x \in \mathbb{N}$, $A \subset \mathbb{N}$, a computable partial functional Φ, numbers $s_i, t_i, u_i \in \mathbb{N}$, and symbols $m_i \in \mathbb{N} \cup \{\uparrow, \downarrow\}$ such that for $i \leq 1$, action for η_i at ρ_i specifies, as a conjunct, $\Phi^{s_i}(A^{t_i} \upharpoonright u_i; x) = m_i$.

DEFINITION 3.1.5. Suppose that $\langle \eta_0, \rho_0 \rangle$ and $\langle \eta_1, \rho_1 \rangle$ share action and that $\rho_0 \subseteq \rho_1$. We say that this shared action is *compatible* if the following conditions hold:

(i) For each set axiom for x and A, if $x \in A^{t_i}$ or $x \notin A^{t_i}$ is supported for η_i at ρ_i, then $t_0 \leq t_1$; and if the set axiom is shared, then either the action for η_0 at ρ_0 is negative, i.e., of the form $x \notin A^{t_0}$, or the action for η_1 at ρ_1 is positive, i.e., of the form $x \in A^{t_1}$.

(ii) For each functional axiom for Φ, A, and x, if the condition, $\Phi^{t_i}(A^{s_i} \upharpoonright u_i; x) = m_i$, reflects action for η_i that is supported at ρ_i, then $s_i \leq t_i$,

$t_0 \leq t_1$, and either action for η_0 at ρ_0 specifies that $m_0 = \uparrow$, or $m_0 = m_1$, or $A^{s_0} \upharpoonright u_0$ and $A^{s_1} \upharpoonright u_1$ are incompatible.

We now prove a lemma listing conditions that imply clause (iv) of Definition 2.10.1. The lemma has three hypotheses about the construction that look somewhat similar on the surface. The first hypothesis states that the construction is faithfully executed, i.e., that the construction succeeds in carrying out the action supported at each node $\rho \subset \Lambda^0$. Thus we allow incompatible axioms, and also allow the removal of elements from sets, even though our aim is to build computably enumerable sets and to avoid declaring incompatible axioms. The third hypothesis stating that shared action is compatible forces the sets constructed to be computably enumerable and the axioms declared to be compatible. The second hypothesis, that the construction admits consistent action is actually superfluous at this point; it is a weaker version of compatibility of shared action. There are really three components to action; the node for which the action is taken, the stage at which the action is taken, and the stage of the approximation to sets use to determine the oracle for axioms. Consistency of action aligns these, comparing action when none of the three components for the first action are longer than the corresponding components for the second action. This condition is generally easy to verify by inspection for a construction. Compatibility of shared action looks at the non-aligned version, and is sometimes harder to prove; when this occurs, the proof generally uses the admission of consistent action. We include this condition here, although it is superfluous, as we will generally first verify it for constructions.

LEMMA 3.1.6. *Suppose that a given construction has the following properties*:

(*i*) *The construction is faithfully executed on* T^0.
(*ii*) *The construction admits consistent action*.
(*iii*) *If* $\langle \eta_0, \rho_0 \rangle$ *and* $\langle \eta_1, \rho_1 \rangle$ *share action, then the action is compatible.*

Then clause (*iv*) *of Definition* 2.10.1 *holds.*

PROOF. Clause (iva) of Definition 2.10.1 follows immediately from (i), and clause (ivb) of Definition 2.10.1 follows from (ii) and (iii). ⊣

When Lemma 3.1.6 is applied, clause (iii) will frequently not be immediately verifiable by inspection. However, there are patterns that are shared by many constructions and imply this condition. In one such pattern, the uses of axioms and the numbers that enter sets are specified as weights of nodes of T^1. We list four conditions describing this pattern, and prove lemmas that can be used to verify Lemma 3.1.6(iii) quickly when this pattern is followed. The first condition places restrictions on the declaration of axioms, the second places restrictions on the entry of numbers into sets, the third deals with the restraint of numbers from entering sets, and the fourth imposes conditions that correct axioms. (The use of the words *newly declared* below means only that this condition is not implied by a previously specified condition.)

DEFINITION 3.1.7 (Standard Parameter Specification). Suppose that $\eta \subset \nu \subseteq \rho$, $\nu^- = \eta$, $\eta_i \subset \rho_i$ for $i \leq 1$, and $\rho_0 \subset \rho_1$. The *standard parameter specification* requires the specification of parameters for sentence decomposition to have the following properties:

(i) If an axiom $\Phi^t(A^s \upharpoonright u; x) \downarrow = m$ is newly declared as action for η supported at ρ, then:
  (a) $u \leq \mathrm{wt}(\lambda(\nu))$,
  (b) $s = \mathrm{wt}(\nu)$,
  (c) $t = \mathrm{wt}(\rho)$, and
  (d) η is the principal derivative of $\mathrm{up}(\eta)$ along ρ.

(ii) If the set axiom $x \in A^s$ is newly declared as action for η supported at ρ, then
  (a) $\mathrm{wt}(\mathrm{up}(\eta)) \leq x < \mathrm{wt}(\lambda(\nu))$,
  (b) $s = \mathrm{wt}(\nu)$,
  (c) $\rho = \nu$, and
  (d) η is validated along ν.

(iii) If the restraint condition $x \notin A^s$ is declared as action for η supported at ρ, then
  (a) $s = \mathrm{wt}(\rho)$,
  (b) $x < \mathrm{wt}(\lambda(\nu))$, and
  (c) $x \notin A^{\mathrm{wt}(\eta)}$.

(iv) If $\langle \eta_0, \rho_0 \rangle$ and $\langle \eta_1, \rho_1 \rangle$ share a functional axiom with oracle A and different values, $\kappa^1 = \lambda(\rho_0) \wedge \lambda(\rho_1)$, then the use of the respective functionals is $\mathrm{wt}(\lambda(\nu_i)) - 1$ where $\nu_i \subseteq \rho_i$ and $\nu_i^- = \eta_i$, and
  (a) $\lambda(\rho_0) | \lambda(\rho_1)$,
  (b) $\kappa^1 \subseteq \mathrm{up}(\eta_0) \wedge \mathrm{up}(\eta_1)$,
  (c) there is a $\kappa \in [\rho_0, \rho_1)$ that is validated along ρ_1 such that $\mathrm{up}(\kappa) = \kappa^1$ and validated action for κ supported at some $\tau \subseteq \rho_1$ places $\mathrm{wt}(\mathrm{up}(\kappa)) \in A$, and
  (d) activated action for κ never places $\mathrm{wt}(\mathrm{up}(\kappa)) \in A$.

We now wish to note that when a construction satisfies these conditions, then restraint imposed to preserve axioms declared for nodes along the current path is always obeyed, and shared action is compatible. The basic idea behind the Restraint Lemma is that any restraint imposed to prevent numbers from entering sets (and so to protect axioms from injury) must be smaller than any numbers placed into the restrained set through supported action at a longer node of T^0. These conditions are what we need, for example, in order to satisfy Friedberg–Mučnik requirements. In addition to the Restraint Lemma (proved as Lemma 8.8.1), we state a Σ_1 Analysis Lemma (proved as Lemma 8.7.1) that describes certain relationships between nodes of T^0 and nodes of T^1. More generally, the lemma can be applied to two trees having consecutive levels, as the theorem relativizes.

LEMMA 3.1.8 (Σ_1 Analysis Lemma). *Let* $\eta \subset \nu \subseteq \rho \subset \tau \in T^0 \cup [T^0]$ *be given such that* $\nu^- = \eta$ *and* η *is* τ*-free. Then the following conditions hold*:

(*i*) *If* $|\tau| = \infty$, *then* $|\{\xi \in (\eta, \tau) : \eta \text{ is } \xi\text{-free}\}| = \infty$.
(*ii*) *If* η *is* ρ*-free, then* $\text{up}(\eta)$ *is* $\lambda(\rho)$*-free.*
(*iii*) $\text{up}(\eta) \subset \lambda(\nu), \lambda(\rho)$.
(*iv*) *If* $\eta^1 \subset \Lambda^1 \in [T^1]$, $\Lambda^0 \in [T^0]$, $\lambda(\Lambda^0) = \Lambda^1$, *and* η^1 *is* Λ^1*-free and has* Π *outcome along* Λ^1, *then* η^1 *has infinitely many* Λ^0*-free derivatives.*
(*v*) *If* $\eta^1 \subset \Lambda^1 \in [T^1]$, $\Lambda^0 \in [T^0]$, $\lambda(\Lambda^0) = \Lambda^1$, *and* η^1 *is* Λ^1*-free and has* Σ *outcome along* Λ^1, *then* η^1 *has a unique* Λ^0*-free derivative* π *with* Π *outcome along* Λ^0, *and* π *is the principal derivative of* η^1 *along* Λ^0.
(*vi*) η *is* ν*-free, and if* η *is the principal derivative of* $\text{up}(\eta)$ *along* ν, *then* $(\lambda(\nu))^- = \text{up}(\eta)$. ⊣

LEMMA 3.1.9 (Restraint Lemma). *Suppose that for a given faithfully executed construction, action for* η *is supported at* ρ *only when* η *is* ρ*-free, sentence decomposition is monotonic, the construction admits consistent action, and clauses* (*ii*) *and* (*iii*) *of Definition* 3.1.7 *are satisfied. Then*:

(*i*) *If clause* (*i*) *of Definition* 3.1.7 *holds, a restraint condition* $x \notin A^s$ *is declared as action for* η *along* ρ, *and* $\text{up}(\eta) \subset \lambda(\rho)$, *then* $x \notin A^{\text{wt}(\rho)}$.
(*ii*) *Shared set action is compatible.*
(*iii*) *If clauses* (*i*) *and* (*iv*) *of Definition* 3.1.7 *hold with the inequality in* (*ia*) *changed to equality and clause* (*iia*) *of Definition* 3.1.7 *is replaced with* $\text{wt}(\text{up}(\eta)) = x$, *then shared functional action is compatible.* ⊣

The conclusions of the Restraint Lemma will still hold when some of the clauses of Definition 3.1.7 are weakened, as will be necessary for some later constructions. We will comment about such weakenings when we encounter constructions requiring them.

3.2. The Friedberg–Mučnik Theorem

In 1944, Post [20] posed the following problem.

Post's Problem: Are there any computably enumerable degrees other than ***0*** and ***0***′?

A negative solution to this problem was obtained more than a decade later by Friedberg [4] and Mučnik [19]. These solutions introduced a new method of proof, the *priority method*, to obtain a pair of incomparable, computably enumerable degrees. A proof of this result, using the Modified Framework Theorem, is presented in this section.

THEOREM 3.2.1 (The Friedberg–Mučnik Theorem). *There are computably enumerable degrees* $\boldsymbol{a}_0$ *and* $\boldsymbol{a}_1$ *such that* $\boldsymbol{a}_0 \mid \boldsymbol{a}_1$.

PROOF. It suffices to construct computably enumerable sets A_0 and A_1 that satisfy the following *incomparability requirements* for all $i \leq 1$ and all computable partial functionals Φ:

$$R_{\Phi,i} : \Phi(A_i) \neq A_{1-i}.$$

Let $\{R_j : j \in \mathbb{N}\}$ be an effective listing of all such requirements.

The idea behind the satisfaction of $R_{\Phi,0}$ (a symmetric description holds for $R_{\Phi,1}$) in the standard construction is to first pick a diagonalization witness x, i.e., an x for which we wish to force $\Phi(A_0; x) \neq A_1(x)$. We then wait to find a number u and a stage s at which $\Phi(A_0 \upharpoonright u; x)[s] \downarrow = 0$ (we say that x is *realized* at s). If no such u and s are found, then we withhold x from A_1 and so satisfy the requirement. If u and s are found, we place x into A_1^{s+1} and try to restrain numbers $< u$ from entering A_0 so that we can preserve the computation $\Phi(A_0 \upharpoonright u; x) = 0$.

Conflicts between requirements arise when a requirement $R_{\Psi,1}$ wishes to put a number $< u$ into A_0. The prioritization of requirements allows an organization of the construction for which, if $R_{\Psi,1}$ has lower priority than $R_{\Phi,0}$, then the diagonalization witness for $R_{\Psi,1}$ is $\geq u$; and if $R_{\Psi,1}$ has higher priority than $R_{\Phi,0}$ and has a diagonalization witness $y < u$, then we reinitialize $R_{\Phi,0}$ and assign a new large diagonalization witness. As there are only finitely many requirements of higher priority than $R_{\Phi,0}$, we can then argue that this requirement has a final diagonalization witness for its satisfaction. The tree argument organizes outcomes, switching when a diagonalization witness is realized.

The trees of strategies approach selects the diagonalization witness x as the weight of the node of T^1 to which the requirement is assigned, and ensures that if the decision to realize x is made at node $\eta \in T^0$, then $u \leq \mathrm{wt}(\lambda(\eta))$. Lemma 3.1.9 (Restraint) ensures that all witnesses newly appointed are larger than any relevant use at the given stage, and switching paths automatically performs initialization. Thus in fact, the trees of strategies proof mimics the standard proof in a different language, and uses framework lemmas to conclude that all requirements are satisfied.

There are several ways to simulate the role of stages in the trees of strategies approach. The easiest technically is to use the weight of a node on T^0 as the stage for evaluating the truth value of the directing sentence. This has a byproduct of evaluating truth values of different sentences at different stages. So we may find an $\eta \in T^0$ and a u such that $\Phi(A_0 \upharpoonright u; x)[\mathrm{wt}(\eta)] \downarrow = 0$ at which we are looking at a lower priority requirement, and place a number $y < u$ into A_0 for the lower priority requirement, thereby injuring the computation. We note that $\Phi(A_0; x) \downarrow = 0$ iff there is a u such that $\Phi(A_0 \upharpoonright u; x)[s] \downarrow = 0$ for infinitely many s iff there is a u such that $\Phi(A_0 \upharpoonright u; x)[s] \downarrow = 0$ for cofinitely many s; so restricting the evaluation of the truth value of the directing sentence for a node of T^1 to the derivatives of that node along a path through T^0 will suffice. This is the approach that we adopt, as it is easiest to use the Restraint Lemma with this approach. A second method is to evaluate the truth value of the above sentence at $\mathrm{wt}(\delta)$ instead of at $\mathrm{wt}(\eta)$, where δ begins the block of T^0 in which η lies. This method will skip stages, but will evaluate all relevant directing sentences at the same stage, so will simplify notation. The standard construction is obtained by letting the stage s be the number of blocks preceding the block in which η lies.

3.2.1. The Basic $R_{\Phi,i}$-Module. This module consists of an initial node α of level 1, and two terminal successors α_0 and α_1 of α; α is activated and has Π outcome along α_0, and is validated and has Σ outcome along α_1. The satisfaction of clauses (i)–(iii) of Definition 2.3.1 follows easily.

Directing sentence S_α: $\exists u \exists \eta \in V_\alpha(\Phi(A_i \upharpoonright u; \mathrm{wt}(\alpha))[\mathrm{wt}(\eta)] \downarrow = 0)$.
Activated action for α at $\rho \supseteq \alpha_0$: $\forall t(\mathrm{wt}(\alpha) \notin A^t_{1-i})$.
Validated action for α at $\rho \supseteq \alpha_1$:

$$\exists r \forall t \geq \mathrm{wt}(\eta)(\mathrm{wt}(\alpha) \in A^r_{1-i} \ \&\ A^t_i \upharpoonright u = A^{\mathrm{wt}(\eta)}_i \upharpoonright u).$$

Note that η and u can be specified as witnesses to the truth of the directing sentence, and are allowed as parameters for validated action. If S_α is true and validated action for α is implemented, then $\Phi(A_i; \mathrm{wt}(\alpha)) \downarrow = 0 \neq 1 = A_{1-i}(\mathrm{wt}(\alpha))$; and if S_α is false, then Lemma 3.1.8(iv) (Σ_1 Analysis) implies that V_α is infinite, so either $\Phi(A_i; \mathrm{wt}(\alpha)) \downarrow \neq 0 = A_{1-i}(\mathrm{wt}(\alpha))$, or $\Phi(A_i; \mathrm{wt}(\alpha)) \uparrow$. In all cases, we see that $\Phi(A_i; \mathrm{wt}(\alpha)) \neq A_{1-i}(\mathrm{wt}(\alpha))$.

3.2.2. T^1-Analysis. The standard initial assignment is followed, and so $R_{\Phi,i} = R_j$ is assigned to each node $\eta^1 \in T^1$ such that $|\eta^1| = j$. Clause (i) of Definition 2.7.2 now follows from Lemma 2.7.4 (Standard Initial Assignment). Hence it suffices to show that all requirements are satisfied.

The standard initial sentence specification for $\eta^1 \in T^1$ is followed, with action supported at $\{\rho^1 \supset \eta^1 : \eta^1 \text{ is } \rho^1\text{-free}\}$. By Lemma 3.1.8(i), ($\Sigma_1$ Analysis), the construction has infinite support on T^1. Clause (i) of Definition 2.10.1 follows from the corresponding analysis for the basic $R_{\Phi,i}$-module.

3.2.3. T^0-Analysis. The standard derived assignment is followed. Clauses (i)–(iii) of Definition 2.7.7 follow from Lemma 2.12.2 (Derived Assignment). Fix $\eta \subset \nu \subseteq \rho \in T^0$ such that $\nu^- = \eta$. Suppose that $\mathrm{up}(\eta) = \eta^1$ is an α-node. The bounds and witnesses for all sentences are as follows: $u = \mathrm{wt}(\lambda(\eta))$, $s = \mathrm{wt}(\eta)$, $t \leq \mathrm{wt}(\rho)$ and $r = \mathrm{wt}(\nu)$. Monotonicity of sentence decomposition follows from (2.8.1) and (2.8.4). Action for η is supported at $\{\rho \supset \eta : \eta \text{ is } \rho\text{-free}\}$. By Lemma 3.1.8(i,ii) ($\Sigma_1$ Analysis), the construction has infinite support on T^0 and support propagates from T^0 to T^1.

3.2.4. The Construction. The construction is executed in accordance with clause (iva) of Definition 2.10.1.

3.2.5. The Verification. Fix the true path $\Lambda^0 \in [T^0]$ determined by the construction, and let $\Lambda^1 = \lambda(\Lambda^0)$. Clauses (i)–(iii) of Definition 3.1.7 are satisfied, so by Lemma 3.1.9(ii) (Restraint), shared set action is compatible. There is no action to declare functional axioms. It easily follows that the construction admits consistent action. Thus we can apply Lemma 3.1.6 to conclude that clause (iv) of Definition 2.10.1 holds. The theorem now follows from Theorem 2.11.5 (Modified Framework). $\dashv$

3.3. A Low Computably Enumerable Degree

A solution to Post's Problem can also be obtained through the construction of a low non-computable computably enumerable degree. The construction for this theorem differs from that of Theorem 3.2.1 (Friedberg–Mučnik) in that there are nodes on all paths through T^1 whose action includes the responsibility to declare the same axiom. Action taken for nodes whose antiderivative is on one path through T^1 may yield a value for the axiom that differs from the value specified by a node whose antiderivative is on an incomparable path through T^1. As the path through T^0 evolves, the current path through T^1 may switch, necessitating a change in value for the axiom. This is resolved by placing a number less than or equal to the use of the axiom into the oracle for the axiom whenever a path is switched, thus allowing a new correct axiom to be defined.

THEOREM 3.3.1 (Low Computably Enumerable Degree Theorem). *There is a computably enumerable degree* $\boldsymbol{a} > 0$ *such that* $\boldsymbol{a}' = \boldsymbol{0}'$.

PROOF. We ensure that $\boldsymbol{a} > \boldsymbol{0}$ by constructing a computably enumerable set A that is not computable. Recall that $\langle \Phi_e : e \in \mathbb{N} \rangle$ is an effective enumeration of all computable partial functionals. We note that for all degrees $\boldsymbol{b}$, all sets $B \subseteq \mathbb{N}$ of degree $\boldsymbol{b}$, and all $e \in \mathbb{N}$, $\{e : \Phi_e(B; e)\downarrow\}$ has degree $\boldsymbol{b}'$. As the jump operator is order-preserving on degrees, we note that $K \leq_T A'$ where K is a fixed computably enumerable set of degree $\boldsymbol{0}'$, the largest computably enumerable degree. Hence we can ensure that A is low by constructing a computably enumerable set D and a computable partial functional Γ such that $\Gamma(D, A) = A'$. Thus it suffices to satisfy the following requirements for all $e \in \mathbb{N}$ and all computable partial functionals Φ:

$$P_\Phi : \Phi(\emptyset) \neq A.$$

$$N_e : \Gamma(D, A; e) = A'(e).$$

Let $\{R_j : j \in \mathbb{N}\}$ be an effective listing of all such requirements. The requirements P_Φ are called *non-computability requirements* and the requirements N_e are called *lowness requirements*.

The satisfaction of a non-computability requirement P_Φ follows a simplified Friedberg–Mučnik strategy, as computations need not be preserved. A follower x for the requirement is chosen, and is withheld from A until a stage s is found such that $\Phi(\emptyset; x)[s] \downarrow = 0$, at which point x is placed into A. For the trees of strategies construction, if P_Φ is assigned to $\eta^1 \in T^1$, then $x = \mathrm{wt}(\eta^1)$, and infinitely many attempts are made to find a convergent computation, at stages $s = \mathrm{wt}(\eta)$ whenever η is a derivative of η^1.

The satisfaction of N_e requires the definition of a computable functional $\Gamma(D, A)$ such that for all $e \in \mathbb{N}$, $\Gamma(D, A; e) = 0$ if $\Phi_e(A; e) \uparrow$, and $\Gamma(D, A; e) = 1$ if $\Phi_e(A; e) \downarrow$. Thus we begin by setting $\Gamma(D, A; e) = 0$ until we find a stage s such that $\Phi_e(A; e)[s] \downarrow$, at which point we wish to reset $\Gamma(D, A; e) = 1$. In order to correct the computation for e, we place the use of the computation $\Gamma(D, A; e) = 0$ into D and restrain A. As the requirement N_e must be assigned along all paths with e unchanged, new types of conflicts arise. Suppose, first, that the requirement is assigned to $\eta^1 \in T^1$ for the first time, and that η^1 is always on the current path. Let η be the initial derivative of η^1 along the current path through T^0, and fix ν on the current path through T^0 such that $\nu^- = \eta$. Then we can choose the use u of the computation $\Gamma(D, A; e) = 0$ to be $\mathrm{wt}(\lambda(\nu)) - 1$, and place $\mathrm{wt}(\eta^1)$ into D in order to correct axioms. Also, infinitely many attempts are made to find a convergent computation, namely at stages $s = \mathrm{wt}(\eta)$ whenever η is a derivative of η^1. However, the current path may, at some point, switch some $\rho^1 \subset \eta^1$ before a convergent computation is found, reassigning N_e to a node ξ^1 with $\mathrm{wt}(\xi^1) > u$. We note that, by (2.8.1), $\mathrm{wt}(\rho^1) \leq u$. In order to carry out our proposed strategy, when ρ^1 is switched, we place $\mathrm{wt}(\rho^1)$ into D if a requirement N_e is assigned to ρ^1, and into A if a requirement P_Φ is assigned to ρ^1.

3.3.1. The Basic P_Φ-Module. This module consists of an initial node α of level 1, and two terminal successors α_0 and α_1 of α; α is activated and has Π outcome along α_0, and is validated and has Σ outcome along α_1. The satisfaction of clauses (i)–(iii) of Definition 2.3.1 follows easily.

Directing sentence S_α: $\exists s(\Phi^s(\emptyset; \mathrm{wt}(\alpha)) \downarrow = 0)$.

Activated action for α at $\rho \supseteq \alpha_0$: $\forall t(\mathrm{wt}(\alpha) \notin A^t)$.

Validated action for α at $\rho \supseteq \alpha_1$: $\exists r(\mathrm{wt}(\alpha) \in A^r)$.

If S_α is true and validated action for α is implemented, then $\Phi(\emptyset; \mathrm{wt}(\alpha)) \downarrow = 0 \neq 1 = A(\mathrm{wt}(\alpha))$; and if S_α is false and activated action for α is implemented, then either $\Phi(\emptyset; \mathrm{wt}(\alpha)) \downarrow \neq 0 = A(\mathrm{wt}(\alpha))$ or $\Phi(\emptyset; \mathrm{wt}(\alpha)) \uparrow$. In either case, we see that $\Phi(\emptyset; \mathrm{wt}(\alpha)) \neq A(\mathrm{wt}(\alpha))$.

3.3.2. The Basic N_e-Module. This module consists of an initial node β of level 1, and two terminal successors β_0 and β_1 of β; β is activated and has Π outcome along β_0, and is validated and has Σ outcome along β_1. The satisfaction of clauses (i)–(iii) of Definition 2.3.1 follows easily.

Directing sentence S_β: $\exists u \exists \eta^0 \in V_\beta(\Phi_e(A \restriction u; e)[\mathrm{wt}(\eta^0)] \downarrow)$.
Activated action for β at $\rho \supseteq \beta_0$:

$$\exists r \forall t \geq r(\Gamma^t(D^t \restriction \mathrm{wt}(\rho), A^t \restriction \mathrm{wt}(\rho); e) = 0 \;\&\; \mathrm{wt}(\beta) \notin D^t).$$

Validated action for β at $\rho \supseteq \beta_1$:

$$\exists r \forall t \geq r(\Gamma^t(D^t \restriction \mathrm{wt}(\rho), A^t \restriction \mathrm{wt}(\rho); e) = 1 \;\&\; \mathrm{wt}(\beta) \in D^r \;\&$$
$$A^t \restriction u = A^{\mathrm{wt}(\eta^0)} \restriction u).$$

The number u and the node η^0 witnessing the truth of the directing sentence S_β may be used as parameters for validated action. If S_β is true and validated action for β is implemented, then $e \in A'$ and $\Gamma(D, A; e) = 1$. And if S_β is false and activated action for β is implemented, then Lemma 3.1.8(iv) (Σ_1 Analysis) implies that V_β is infinite, so $e \notin A'$ and $\Gamma(D, A; e) = 0$. In either case, N_e is satisfied.

3.3.3. T^1-Analysis. The standard initial assignment is followed, and so R_j is assigned to each node $\eta^1 \in T^1$ such that $|\eta^1| = j$. Clause (i) of Definition 2.7.2 now follows from Lemma 2.7.4 (Standard Initial Assignment). Hence it suffices to show that all requirements are satisfied.

The standard initial sentence specification for $\eta^1 \in T^1$ is followed, with action supported at $\{\rho^1 \supset \eta^1 : \eta^1 \text{ is } \rho^1\text{-free}\}$. By Lemma 3.1.8(i) ($\Sigma_1$ Analysis), the construction has infinite support on T^1. Clause (i) of Definition 2.10.1 follows from the corresponding analysis for the basic modules.

3.3.4. T^0-Analysis. The standard derived assignment is followed. Clauses (i)–(iii) of Definition 2.7.7 follow from Lemma 2.12.2 (Derived Assignment). Fix $\eta \subset \nu \subseteq \rho \in T^0$ such that $\nu^- = \eta$. The witnesses and bounds for all sentences are as follows: $u = \mathrm{wt}(\lambda(\eta))$, $s = \mathrm{wt}(\eta)$, $t \leq \mathrm{wt}(\rho)$ and $r = \mathrm{wt}(\nu)$. Monotonicity of sentence decomposition follows from (2.8.1) and (2.8.4). Action for η is supported at $\{\rho \supset \eta : \eta \text{ is } \rho\text{-free}\}$. By Lemma 3.1.8(i,ii) ($\Sigma_1$ Analysis), the construction has infinite support on T^0 and support propagates from T^0 to T^1.

3.3.5. The Construction. The construction is executed in accordance with clause (iva) of Definition 2.10.1.

3.3.6. The Verification. Fix the true path $\Lambda^0 \in [T^0]$ determined by the construction, and let $\Lambda^1 = \lambda(\Lambda^0)$. It is easily checked that the construction admits consistent action. If a functional axiom for η is newly declared at ν with use $u+1$, then η is ν-free, so by Lemma 3.1.8(vi) (Σ_1 Analysis), $\nu^- = \eta$; hence by Lemma 3.1.8(ii,vi) (Σ_1 Analysis), $u = \mathrm{wt}(\lambda(\nu))$ and $(\lambda(\nu))^- = \mathrm{up}(\eta)$. It is now easy to see that clauses (i)–(iv) of Definition 3.1.7 are satisfied. Thus by Lemma 3.1.9(ii,iii) (Restraint), shared set action is compatible and shared functional action is compatible. Thus we can apply Lemma 3.1.6 to conclude that clause (iv) of Definition 2.10.1 holds. The theorem now follows from Theorem 2.11.5 (Modified Framework). ⊣

3.4. A Properly d-c.e. Degree

Not all constructions have basic modules consisting of a single non-terminal node. The theorem proved in this section was chosen as it is a simple example of a theorem whose proof requires a larger basic module.

The construction we give violates clauses (ii) and (iii) of Definition 3.1.7 (Standard Parameter Specification). However, those conditions are stronger than is necessary to prove a restraint lemma. We introduce weakened versions of these conditions below, together with the required replacement for Lemma 3.1.9 (Restraint). This lemma is proved as Lemma 8.8.2.

DEFINITION 3.4.1 (Delayed Action Specification). Suppose that $\eta \subset \nu \subseteq \rho$ and $\nu^- = \eta$. The *delayed action specification* requires a construction to satisfy the following conditions:

(i) If the set axiom $x \in A^s$ is newly declared as action for η supported at ρ, then there are $\xi \subseteq \eta$ and $\xi^1 = \text{up}(\xi) \subseteq \text{up}(\eta) = \eta^1$ such that
  (a) $x = \text{wt}(\xi^1)$,
  (b) $s = \text{wt}(\nu)$,
  (c) $\rho = \nu$, and
  (d) ξ and η are both validated along ν.

(ii) If the restraint condition $x \notin A^s$ is declared as action for η supported at ρ, then there are $\xi \subseteq \eta$ and $\xi^1 = \text{up}(\xi) \subseteq \text{up}(\eta) = \eta^1$ such that
  (a) $s = \text{wt}(\rho)$,
  (b) $x < \text{wt}(\lambda(\nu))$, and
  (c) $x \notin A^{\text{wt}(\xi)}$ and either $x \notin A^{\text{wt}(\eta)}$, or $x = \text{wt}(\xi^1)$ and there do not exist $\sigma^1 \in [\xi^1, \eta^1]$ and $\tau^1 \in (\sigma^1, \rho^1]$ such that action for σ^1 supported at τ^1 specifies that $x \in A$.

LEMMA 3.4.2 (Delayed Set Action Lemma). *Suppose that for a given faithfully executed construction, action for η is supported at ρ only when η is ρ-free, sentence decomposition is monotonic, the construction admits consistent action, and clauses (i) and (ii) of Definition* 3.4.1 *are satisfied. Then shared set action is compatible.*

The class of computably enumerable sets is a natural class of sets to study, as these sets are generated as output sets of computer programs operating without reference to an oracle. This class is closed under union and intersection, but not under complementation or set difference. When studying degrees, the complement of a set A has the same degree as that of A; it is natural to next generate a hierarchy, looking at sets that can be generated from computably enumerable sets by boolean operations. The simplest such class after the class of computably enumerable sets is the class of sets that are differences of computably enumerable sets, and it is natural to ask whether the degree of each such set is also the degree of a computably enumerable set. Our next theorem shows that this is not the case, i.e., that new degrees are produced by going to this larger class of sets.

DEFINITION 3.4.3. $D \subseteq \mathbb{N}$ is *d-c.e.* (*the difference of computably enumerable sets*) if there are computably enumerable sets A and B such that $D = A - B$. A degree $\boldsymbol{d}$ is *d-c.e.* if $\boldsymbol{d}$ is the degree of a d-c.e. set. A d-c.e. degree $\boldsymbol{d}$ is *properly d-c.e.* if $\boldsymbol{d}$ does not contain a computably enumerable set.

Properly d-c.e. degrees were first constructed by Cooper [3], but the proof that we present is due to Lachlan and is substantially simpler.

THEOREM 3.4.4. *There is a properly d-c.e. degree.*

PROOF. We construct computably enumerable sets A and B for which the degree of $D = A - B$ is a properly d-c.e. degree. For each computably enumerable set W and each pair $\langle\Phi, \Psi\rangle$ of computable partial functionals, we establish the requirement

$$R_{\Phi,\Psi,W} : \Phi(W) \neq D \text{ or } \Psi(D) \neq W,$$

and all requirements are organized into an effective list $\{R_i : i \in \mathbb{N}\}$.

The strategy for satisfying the requirement $R_{\Phi,\Psi,W}$ is as follows. A number x ($x = \mathrm{wt}(\alpha)$ in the basic module) is designated, and the construction searches for a stage s ($s = \mathrm{wt}(\gamma^-)$ in the module) and numbers u and v such that $\Phi(W \restriction u; x)[s] \downarrow = 0$ and D pins down $W^s \restriction u$, i.e., $\Psi(D \restriction v)[s] \restriction u = W^s \restriction u$. While the search is unsuccessful, $D(x) = 0$. If s is found, then x is placed into A, thus temporarily setting $D(x) = 1$, and all other numbers $< v$ are restrained from entering or leaving D. If $W \restriction u = W^s \restriction u$, then $\Phi(W; x) \downarrow = 0 \neq 1 = D(x)$. The construction now searches for a stage $t > s$ such that $W^t \restriction u \neq W^s \restriction u$. If such a t is found, then x is placed into B, thereby resetting $D \restriction v = D^s \restriction v$, and thus forcing $\Psi(D; z) \neq W(z)$ for some $z < u$. Lemma 3.4.2 allows us to show that the iterated trees of strategies approach automatically resolves conflicts between requirements; such conflicts arise only when one requirement wishes to place an element into D or remove an element from D, and another wishes to restrain D in order to preserve a computation.

3.4.1. The Basic $R_{\Phi,\Psi,W}$-Module. This module has two non-terminal nodes α and α_1 of level 1, and three terminal nodes α_0, α_{10} and α_{11}. According to the convention, α will be activated (validated, resp.) and have Π (Σ, resp.)

outcome along α_0 (α_1, resp.); and α_1 will be activated (validated, resp.) and have Π (Σ, resp.) outcome along α_{10} (α_{11}, resp.). Note that clauses (i)–(iii) of Definition 2.3.1 are satisfied.

Directing sentence S_α:

$$\exists u \exists v \exists \eta \in V_\alpha(\Phi(W \restriction u; \mathrm{wt}(\alpha))[\mathrm{wt}(\eta)] \downarrow = 0 \; \& \; \Psi(D \restriction v)[\mathrm{wt}(\eta)] \restriction u = W^{\mathrm{wt}(\eta)} \restriction u).$$

Activated action for α at $\rho \supseteq \alpha_0$: $\forall t(\mathrm{wt}(\alpha) \notin A^t \cup B^t)$.
Validated action for α at $\rho \supseteq \alpha_1$:

$$\exists r(\mathrm{wt}(\alpha) \in A^r) \; \& \; \forall x < v \forall t \geq \mathrm{wt}(\eta)(x \neq \mathrm{wt}(\alpha) \rightarrow A^t(x) = A^{\mathrm{wt}(\eta)}(x) \; \& \; B^t(x) = B^{\mathrm{wt}(\eta)}(x)).$$

We note that the witnesses η, u, and v to the validity of S_α may be used as parameters for validated action for α and in the directing sentence and action sentences for α_1. Let $\alpha_1 = \alpha^\frown\langle\gamma\rangle$.

Directing sentence S_{α_1}: $\exists z < u \exists s > \mathrm{wt}(\eta)(z \notin W^{\mathrm{wt}(\eta)} \; \& \; z \in W^s)$.
Activated action for α_1 at $\rho \supseteq \alpha_{10}$: $\forall t(\mathrm{wt}(\alpha) \notin B^t)$.
Validated action for α_1 at $\rho \supseteq \alpha_{11}$: $\exists r(\mathrm{wt}(\alpha) \in B^r)$.

If the true path passes through α_0, then as Lemma 3.1.8(iv)(Σ_1 Analysis) implies that V_α is infinite, either $\Phi(W; \mathrm{wt}(\alpha)) \uparrow$; or $\Phi(W; \mathrm{wt}(\alpha)) \downarrow \neq 0 = D(\mathrm{wt}(\alpha))$; or $\Phi(W; \mathrm{wt}(\alpha)) \downarrow = 0$ with some use u and either $\Psi(D; z) \uparrow$ for some $z < u$, or $\Psi(D; z) \downarrow \neq W(z)$ for some $z < u$. If the true path passes through α_{10}, then $\Phi(W; \mathrm{wt}(\alpha)) \downarrow = 0 \neq 1 = D(\mathrm{wt}(\alpha))$. And if the true path passes through α_{11}, then $\Psi(D; z) \downarrow \neq W(z)$ for some $z < u$.

3.4.2. T^1-Analysis. The standard initial assignment is followed. Clause (i) of Definition 2.7.2 follows from Lemma 2.7.4 (Standard Initial Assignment). Hence it suffices to show that all requirements are satisfied.

The standard initial sentence specification for $\eta^1 \in T^1$ is followed, with action supported at $\{\rho^1 \supset \eta^1 : \eta^1 \text{ is } \rho^1\text{-free}\}$. By Lemma 3.1.8(i), ($\Sigma_1$ Analysis), the construction has infinite support on T^1. Clause (i) of Definition 2.10.1 follows from the corresponding analysis for the basic module.

3.4.3. T^0-Analysis. The standard derived assignment is followed. Clauses (i)–(iii) of Definition 2.7.7 follow from Lemma 2.12.2 (Derived Assignment).

Fix $\eta \subset \nu \subseteq \rho \in T^0$ such that $\nu^- = \eta$. The witnesses and bounds for the sentences for η are as follows: $u \leq \mathrm{wt}(\lambda(\eta))$, $v = \mathrm{wt}(\lambda(\eta))$, $t \leq \mathrm{wt}(\rho)$ and $r = \mathrm{wt}(\nu)$. Monotonicity of sentence decomposition follows from (2.8.1) and (2.8.4). Action for η is supported at $\{\rho \supset \eta : \eta \text{ is } \rho\text{-free}\}$. By Lemma 3.1.8(i,ii), (Σ_1 Analysis), the construction has infinite support on T^0 and support propagates from T^0 to T^1.

3.4.4. The Construction. The construction is executed in accordance with clause (iva) of Definition 2.10.1.

3.4.5. The Verification. Fix the true path $\Lambda^0 \in [T^0]$ determined by the construction, and let $\Lambda^1 = \lambda(\Lambda^0)$. It is easily checked that the construction admits consistent action. Clauses (i) and (ii) of Definition 3.4.1 are satisfied, so by Lemma 3.4.2 (Delayed Set Action), shared set action is compatible. There is no functional action. Thus Lemma 3.1.6 allows us to conclude that clause (iv) of Definition 2.10.1 holds. The theorem now follows from Theorem 2.11.5 (Modified Framework). $\dashv$

3.5. Summary

In this chapter, Player II was responsible for constructing all the sets, and the functionals defined by Player II had a set of degree $\boldsymbol{0}'$ as their oracle. As positive set action to place a number x into a fixed set A was localized within a single basic module on T^1, properties of the weight function, together with the ability to keep modules on T^1 intact and internally consistent sufficed to ensure that shared set action was compatible. Shared functional action occurred only along incomparable paths through T^1, and we were able to ensure correction whenever the path on T^1 was switched. More sophisticated techniques will be required for constructions at a higher level.

Chapter 4

Δ_2 CONSTRUCTIONS

Δ_2 constructions are the simplest priority constructions in which Player I controls a set W. Such constructions take place on three trees, T^0, T^1, and T^2. In a typical situation, W is non-computable and the success of the construction requires the entry of numbers into W at a stage at which such numbers are useful to the construction. To show that this must happen, the construction will try to compute more and more of W at prescribed sets of stages, so as W is not computable, these attempts must fail; and the failure will produce numbers that enter W when they are useful.

We discuss general Δ_2 constructions in Section 4.1, stating a lemma describing relationships between nodes of that level. The lemma is applied to prove an upward cone avoidance theorem in Section 4.2, the Sacks Splitting Theorem in Section 4.3, basic facts about backtracking are presented in Section 4.4 and a permitting construction is carried out in Section 4.5.

4.1. The Δ_2 Level

Unlike Σ_1 constructions, Δ_2 constructions will have level 2 requirements. However, these requirements will be of a very special nature, making these constructions somewhat similar to those at the Σ_1 level. Prior to the existence of the classification of priority arguments using the arithmetical hierarchy, a differentiation was made between *finite injury* and *infinite injury* priority constructions only, and Δ_2 constructions were classified as finite injury. But even at that time, a distinction was made between the two types of finite injury constructions.

Δ_2 constructions are constructions having requirements of level 2 but no higher, but such that all requirements of level 2 are *degenerate*; by this we mean that when such a requirement is assigned to a level 2 node $\eta^2 \in T^2$, and $\tau^2 \supset \eta^2$ is such that $(\tau^2)^- = \eta^2$ and η^2 has Π outcome along τ^2, then no requirement is assigned to τ^2 or any of its successors. (Recall that the standard initial assignment implements the situation just described.) This causes all derivatives of η^2 on T^1 to lie in intervals of Λ^1, all of whose nodes are derivatives of η^2 or its successors. If the interval is finite, then η^2 will have Σ outcome along $\lambda(\Lambda^1)$, and the satisfaction of the requirement will be finitary; and if the interval is infinite, then η^2 will have Π outcome along $\lambda(\Lambda^1)$, and only finitely nodes of T^2 will have derivatives along $\lambda(\Lambda^1)$; in this case, we will show that the theorem follows from the directing sentences and action for η^2 and its predecessors along the true path. For many constructions, the latter situation will provide an effective computation of a set W. The corresponding theorem will have the statement, "W is computable or ...".

The conditions of Lemma 3.1.8 (Σ_1 Analysis) hold for all constructions. However, additional properties of the Δ_2 level are needed to analyze Δ_2 constructions. The Δ_2 Analysis Lemma (proved as Lemma 8.7.2) specifies these properties.

LEMMA 4.1.1 (Δ_2 Analysis Lemma). *For a Δ_2 construction, fix $\eta^2 \in T^2$ such that* $\text{lev}(\eta^2) = 2$. *Suppose that $\xi^1, \eta^1 \in T^1$ and* $\text{up}(\xi^1) = \text{up}(\eta^1) = \eta^2$. *Then*:

(*i*) *ξ^1 and η^1 are comparable nodes of T^1, say $\xi^1 \subset \eta^1$, and all nodes in $[\xi^1, \eta^1]$ are derived from η^2.*

(*ii*) *If $\eta^1 \subset \lambda(\eta)$ is the longest derivative of η^2 that is $\subset \lambda(\eta)$, then either η^1 has Π outcome along $\lambda(\eta)$ and is the only $\lambda(\eta)$-free derivative of η^2, or η^1 has Σ outcome along $\lambda(\eta)$ and all derivatives of η^2 along $\lambda(\eta)$ are $\lambda(\eta)$-free.*

4.2. Avoiding Upward Cones

Our first, and simplest Δ_2 construction is that of a non-computable, computably enumerable set A such that $W \not\leq_T A$, where W is a given non-computable, computably enumerable set. We ensure that A is non-computable through the use of the non-computability requirements introduced in Section 3.3. Our other *preservation requirements* ensure that, for all computable partial functionals Φ, $\Phi(A) \neq W$. While these requirements resemble Friedberg–Mučnik requirements, this resemblance is superficial, as we have no control over the entry of numbers into W. More precisely, if we declare an axiom $\Phi(A; x)[s] \neq W^s(x)$ and $x \in W^t - W^s$, then our attempt at satisfying the requirement through the use of the witness x has failed. Instead, we adopt a strategy that imposes restraint on A; we define a computable partial functional Γ, and declare axioms $\Gamma(\emptyset; x) = \Phi(A; x)$ whenever $\Phi(A) \upharpoonright x+1 = W \upharpoonright x+1$. Thus the failure to satisfy $\Phi(A) \neq W$ results in an effective computation of W. In the setting of Δ_2 constructions, the theorem actually proved is that either W is computable, or there is a non-computable, computably enumerable set A such that $W \not\leq_T A$.

As infinitely many axioms must be declared, preservation requirements are level 2 requirements. These requirements may need to act at infinitely many nodes along a path through T^1, so we need a scheme that will allow this to happen while allowing other requirements that need to be satisfied to act as well. Non-computability requirements of lower priority will place numbers into A, allowing W to change; as $\Gamma(\emptyset)$ has no ability to correct computations, the injury to the preservation requirement would be fatal. In a Δ_2 construction, we can prevent lower priority requirements from acting, and such action is not necessary as long as we satisfy $\Gamma(\emptyset) = \Phi(A) = W$; for then W will be computable, which is one of the alternatives provided by the theorem. And once we have an x such that $\Gamma(\emptyset; x)\downarrow = \Phi(A; x)\downarrow \neq W(x)$, we have satisfied this requirement and no further axioms are needed. Restraint on A can then be imposed, and lower priority requirements are initialized and can then begin to act.

THEOREM 4.2.1 (Upward Cone Avoidance Theorem). *For every computably enumerable set W, either W is computable, or there is a non-computable, computably enumerable set A such that $W \not\leq_T A$.*

PROOF. We attempt to satisfy the following requirements for all computable partial functionals Ψ and Φ:

$$P_\Phi : \Phi(\emptyset) \neq A.$$
$$N_\Phi : \Phi(A) = W \Rightarrow \exists\Gamma(\Gamma(\emptyset) = \Phi(A)).$$

Let $\{R_j : j \in \mathbb{N}\}$ be an effective listing of all requirements.

4.2.1. The Basic P_Φ-Module. This is the module introduced in Section 3.3, so is not presented here; its initial node is an α-node.

4.2.2. The Basic N_Φ-Module. The directing sentence for $\eta^2 \in T^2$ working for this requirement checks to see if $\Phi(A) = W$. At a derivative $\eta^1 \in T^1$, the check is just for $\Phi(A)$ and W to agree through $\mathrm{wt}(\eta^1)$. At a derivative $\eta \in T^0$ of η^1, we check to see if the desired agreement occurs at stage $\mathrm{wt}(\eta)$. The directing sentence checks just at nodes in V_{η^2} and V_{η^1}, and this will suffice if, in fact, $\Phi(A) = W$. If we find the desired agreement, then action will be to restrain A and thereby preserve the computations $\Phi(A; x)$ for $x \leq \mathrm{wt}(\eta^1)$, and to define axioms $\Gamma(\emptyset; x) = \Phi(A; x)$ for such x. As all derivatives of η^2 lie in an interval of T^1, such restraint will ensure that shared functional axioms for Γ are compatible. Should W change, then the A restraint will cause $\Phi(A)$ and W to disagree, so the requirement will be satisfied finitarily.

The module consists of an initial node β of level 2, with two terminal immediate successors β_0 and β_1; β is activated and has Σ outcome along β_0, and is validated and has Π outcome along β_1. The satisfaction of of clauses (i)–(iii) of Definition 2.3.1 follows easily. There is no activated action for β. We note that the witnesses u_{β^1} and $\beta^0_{\beta^1}$ provided by the truth of the directing sentence for β^1 can be used in the validated action sentence for the same β^1.

Directing sentence S_β:

$$\forall \beta^1 \in V_\beta \exists u \exists \beta^0 \in V_{\beta^1} \forall y \leq \mathrm{wt}(\beta^1)(\Phi(A \upharpoonright u; y)[\mathrm{wt}(\beta^0)] \downarrow = W^{\mathrm{wt}(\beta^0)}(y)).$$

Validated action for β at $\rho \supseteq \beta_1$:

$$\forall \beta^1 \in V_\beta \exists r \forall t \geq \mathrm{wt}(\beta^0_{\beta^1}) \forall y \leq \mathrm{wt}(\beta^1)$$
$$(\Gamma^r_\beta(\emptyset; y) \downarrow = \Phi(A \upharpoonright u_{\beta^1}; y)[\mathrm{wt}(\beta^0_{\beta^1})] \;\&\; A^t \upharpoonright u_{\beta^1} = A^{\mathrm{wt}(\beta^0_{\beta^1})} \upharpoonright u_{\beta^1}).$$

Suppose first that $\beta \subset \Lambda^2$, S_β is true, and validated action for β is followed. By Lemma 3.1.8(iv) (Σ_1 Analysis) relativized to T^1, β has infinitely many Λ^1-free derivatives. Validated action now ensures that $\Gamma(\emptyset) = \Phi(A)$ and both are total, and the restraint condition of validated action ensures that $\Phi(A) = W$. Now suppose that S_β is false, and β has Σ outcome along Λ^2. Then by Lemma 3.1.8(v) (Σ_1 Analysis) relativized to T^1, β has a unique Λ^1-free derivative β^1 with Π outcome, so we cannot find u and β^0 for β^1 as specified by S_β. By Lemma 3.1.8(iv) (Σ_1 Analysis), β^1 has infinitely many Λ^0-free derivatives. Hence there must be a $y \leq \mathrm{wt}(\beta^1)$ for which $\Phi(A; y) \neq W(y)$. We have thus shown that N_Φ is satisfied.

4.2.3. T^2-Analysis. The standard initial assignment is followed. Clause (i) of Definition 2.7.2 follows from Lemma 2.7.4 (Standard Initial Assignment). Hence it suffices to show that all requirements along the true path $\Lambda^2 \in [T^2]$ are satisfied.

The standard initial sentence specification for $\eta^2 \in T^2$ is followed, with action supported at $\{\rho^2 \supset \eta^2 : \eta^2 \text{ is } \rho^2\text{-free}\}$. Clause (i) of Definition 2.10.1 follows from the corresponding analysis for the basic modules.

4.2.4. T^1-Analysis. The standard derived assignment is followed. Clauses (i)–(iii) of Definition 2.7.7 follow from Lemma 2.12.2 (Standard Derived Assignment).

If $\eta^1 \in T^1$ is a β-node, then sentence decomposition bounds $\beta^1 \subseteq \eta^1$. Monotonicity of sentence decomposition follows easily. Action for $\eta^1 \in T^1$ is supported at $\{\rho^1 \supset \eta^1 : \eta^1 \text{ is } \rho^1\text{-free}\}$. By Lemma 3.1.8(i,ii) ($\Sigma_1$ Analysis), the construction has infinite support on T^1, and support propagates from T^1 to T^2.

4.2.5. T^0-Analysis. The standard derived assignment is followed. Clauses (i)–(iii) of Definition 2.7.7 follow from Lemma 2.12.2 (Standard Derived Assignment).

Sentence decomposition for α-nodes is as described in Section 3.3. Fix a β-node $\eta \subset \nu \subseteq \rho \in T^0$ such that $\nu^- = \eta$, and let $\eta^1 = \mathrm{up}(\eta)$. The bounds for the sentences are: $\beta^0 \subseteq \eta$, $u = \mathrm{wt}(\lambda(\eta))$, $r = \mathrm{wt}(\nu)$ and $t \leq \mathrm{wt}(\rho)$. Monotonicity of sentence decomposition follows from (2.8.1) and (2.8.4). Action for η is supported at $\{\rho \supset \eta : \eta \text{ is } \rho\text{-free}\}$. By Lemma 3.1.8(i,ii) ($\Sigma_1$ Analysis), the construction has infinite support on T^0, and support propagates from T^0 to T^1.

4.2.6. The Construction. The construction is executed in accordance with clause (iva) of Definition 2.10.1.

4.2.7. The Verification. Fix the true path $\Lambda^0 \in [T^0]$ determined by the construction, and let $\Lambda^{k+1} = \lambda(\Lambda^k)$ for $k \leq 1$. It is easily checked that the construction admits consistent action. Clauses (i)–(iii) of Definition 3.1.7 are easily seen to be satisfied, so it follows from Lemma 3.1.9(ii) (Restraint) that shared set action is compatible. By Lemma 4.1.1 (Δ_2 Analysis), all axioms for Γ are declared along a single path through T^1, and whenever an axiom is newly declared at $\rho \subset \Lambda^0$, action for all previously declared axioms is supported at ρ. Furthermore, Lemma 3.1.9(ii) (Restraint) implies that the value for an axiom never changes and so that shared functional action is compatible. Hence we can apply Lemma 3.1.6 to conclude that clause (iv) of Definition 2.10.1 holds. The theorem now follows from Theorem 2.11.5 (Modified Framework). ⊣

4.3. The Sacks Splitting Theorem

The Sacks Splitting Theorem [21] is not a theorem strictly about $\mathcal{R}$, although it has a corollary that specifies a property of $\mathcal{R}$. For every number x, there is a requirement whose validated action places x into a set, so we cannot rely on weight computations to automatically manage conflicts between such requirements and requirements imposing restraints on sets. These conflicts are resolved in advance by a careful assignment of requirements to nodes of T^2, so that when a restraint $x \notin A$ is imposed by a node, successors of that node will choose a set other than A as the target set for x. Thus clause (iia) of Definition 3.1.7 will not be satisfied, so we will need a new version of the Restraint Lemma (proved as Lemma 8.8.3).

LEMMA 4.3.1 (Splitting Restraint Lemma). *Suppose that we are given a Δ_2 construction satisfying clauses (i), (iib,c,d) and (iii) of Definition 3.1.7, as well as the following condition:*

(*i*) *If $\eta^{2\frown}\langle\gamma^1\rangle = \nu^2 \subseteq \sigma^2 \in T^2$, η^2 has Σ outcome along ν^2, action for η^2 along $\rho^2 \supseteq \nu^2$ specifies a restraint condition for the set A, and action for σ^2 along $\tau^2 \supseteq \nu^2$ specifies that a number x be placed in the same set A, then $x \geq \mathrm{wt}(\gamma^1)$.*

Suppose also that the construction is faithfully executed, action for η is supported at ρ only when η is ρ-free, sentence decomposition is monotonic, and the construction admits consistent action. Then conclusions (i) and (ii) of Lemma 3.1.9 (Restraint) hold. Furthermore, if hypothesis (iii) of Lemma 3.1.9 holds for the functional Δ, then shared functional action for Δ is compatible. ⊣

THEOREM 4.3.2 (Sacks Splitting Theorem). *Let W be a computably enumerable set. Then there are computably enumerable sets A_0 and A_1 such that if W is not computable, then $W = A_0 \cup A_1$, $A_0 \cap A_1 = \emptyset$, and $W \not\leq_T A_0, A_1$.*

The following corollary follows easily from Theorem 4.3.2.

COROLLARY 4.3.3. *For every computably enumerable degree $\mathbf{a} > \mathbf{0}$, there are computably enumerable degrees $\mathbf{a_0}, \mathbf{a_1} < \mathbf{a}$ such that $\mathbf{a_0} \cup \mathbf{a_1} = \mathbf{a}$.*

PROOF. (of Theorem 4.3.2) Note that if W is not computable, then A_0 and A_1 cannot be computable. For since A_0 and A_1 partition W, $W \equiv_T A_0 \oplus A_1$, so if, e.g., A_0 is computable, then $A_1 \equiv_T W$.

For each $x \in \mathbb{N}$, we establish the requirement

$$P_x : (x \in W \Rightarrow x \in A_0 \cup A_1 \;\&\; x \notin A_0 \cap A_1) \;\&\; (x \notin W \Rightarrow x \notin A_0 \cup A_1);$$

and for each $i \leq 1$ and computable partial functional Φ, we establish the requirement

$$N_{\Phi,i} : \Phi(A_i) = W \Rightarrow \exists\Delta(\Delta(\emptyset) = \Phi(A_i)).$$

All requirements are organized into an effective list $\{R_i : i \in \mathbb{N}\}$. The requirements P_x ensure that A_0 and A_1 partition W, and the requirements $N_{\Phi,i}$ are the preservation requirements of the previous section. A Δ_2 construction will allow us to avoid conflicts between requirements by judiciously choosing the set A_i into which we place an x just entering W; only one requirement $N_{\Phi,j}$ will be imposing a restraint on x, and we choose $i = 1 - j$.

4.3.1. The Basic P_x-Module. This module consists of an initial node α of level 1, and two terminal successors α_0 and α_1; α is activated and has Π outcome along α_0 and is validated and has Σ outcome along α_1. Clauses (i)–(iii) of Definition 2.3.1 follow easily. The validated action sentence is a disjunction, so is nondeterministic; however, the choice of disjuncts will be made when the requirement is assigned to a node of T^2, so the resulting construction will be deterministic.

Directing sentence S_α: $\exists s(x \in W^s)$.
Activated action for α at $\rho \supset \alpha_0$: $\forall t(x \notin A_0^t \cup A_1^t)$.
Validated action for α at $\rho \supset \alpha_1$: $\exists i \leq 1 \exists r \forall t \geq r(x \in A_i^t - A_{1-i}^t)$.

It is clear that the requirement P_x is satisfied if either S_α is true and validated action for α is followed, or S_α is false and activated action for α is followed.

4.3.2. The Basic $N_{\Phi,i}$-Module. This module is the strategy for preservation requirements introduced in the last section, so will not be repeated here. Nodes of T^2 to which such requirements are assigned are called β-nodes.

The proof of the satisfaction of these requirements will utilize Lemma 4.3.1 (Splitting Restraint), so we will have to ensure that clause (i) of that lemma is satisfied. The initial assignment of requirements to nodes of T^2 is non-standard, and designed to obtain this property by assigning requirements P_x to nodes of T^2 in blocks and specifying the target set for x appropriately.

4.3.3. T^2-Analysis. The first requirement of the form $N_{\Phi,i}$ is assigned to $\emptyset$. Suppose that $\eta^2 \in T^2$ is given and requirements have been assigned to all $\xi^2 \subset \eta^2$ but no requirement has yet been assigned to $\eta^2 \neq \emptyset$. If there is a β-node $\xi^2 \subset \eta^2$ that has Π outcome along η^2, then no requirement is assigned to η^2. If there is a β-node $\xi^2 \subset \eta^2$ that has Σ outcome along η^2, say $\xi^{2\frown}\langle\gamma^1\rangle \subseteq \eta^2$ with $\gamma^1 \in T^1$, and there is an $x < \mathrm{wt}(\gamma^1)$ such that P_x has not yet been assigned to any $\sigma^2 \subset \eta^2$, then for the least such x, P_x is assigned to η^2. If the requirement $N_{\Phi,i}$ is assigned to ξ^2, then the disjunct chosen for validated action is $\exists r \forall t \geq r(x \in A^t_{1-i} - A^t_i)$. (This step will ensure the preservation of all axioms discovered for S_β whenever β lies along the true path of T^2.) Otherwise, the first requirement of our list that has not yet been assigned to any $\xi^2 \subset \eta^2$ is assigned to η^2, with validated action targeting x for A_0 if the requirement is of the form P_x. Clause (i) of Definition 2.7.2 follows, as W is computable whenever we follow a validated extension of a β-node of T^2. Furthermore, the assignment of requirements is defined in accordance with clause (i) of Lemma 4.3.1 (Splitting Restraint), so it straightforwardly follows that this condition is satisfied.

The standard initial sentence specification for $\eta^2 \in T^2$ is followed, with action supported at $\{\rho^2 \supset \eta^2 : \eta^2 \text{ is } \rho^2\text{-free}\}$. Clause (i) of Definition 2.10.1 follows from the corresponding analysis for the basic modules. Hence it suffices to show that all requirements are satisfied.

4.3.4. T^1-Analysis. The standard derived assignment is followed. Clauses (i)–(iii) of Definition 2.7.7 follow from Lemma 2.12.2 (Standard Derived Assignment).

Sentence decomposition for β nodes is as in the preceding section, and will be described for α-nodes when we reach T^0. Action for $\eta^1 \in T^1$ is supported at $\{\rho^1 \supset \eta^1 : \eta^1 \text{ is } \rho^1\text{-free}\}$. By Lemma 3.1.8(i,ii) ($\Sigma_1$ Analysis), the construction has infinite support on T^1, and support propagates from T^1 to T^2.

4.3.5. T^0-Analysis. The standard derived assignment is followed. Clauses (i)–(iii) of Definition 2.7.7 follow from Lemma 2.12.2 (Standard Derived Assignment).

Sentence decomposition for α-nodes η is as follows. Fix $\rho \supseteq \nu \supset \eta$ such that $\nu^- = \eta$. The bounds and witnesses for α-nodes are as follows: $s = \mathrm{wt}(\eta)$, $t \leq \mathrm{wt}(\rho)$ and $r = \mathrm{wt}(\nu)$. Sentence decomposition for β-nodes is as in Section 4.2. Monotonicity of sentence decomposition follows from (2.8.1) and (2.8.4). Action for η is supported at $\{\rho \supset \eta : \eta \text{ is } \rho\text{-free}\}$. By Lemma 3.1.8(i,ii) ($\Sigma_1$ Analysis), the construction has infinite support on T^0, and support propagates from T^0 to T^1.

4.3.6. The Construction. The construction is executed in accordance with clause (iva) of Definition 2.10.1.

4.3.7. The Verification. Fix the path $\Lambda^0 \in [T^0]$ determined by the construction, and let $\Lambda^{k+1} = \lambda(\Lambda^k)$ for $k \leq 1$. It is easily checked that the construction admits consistent action. We note that clauses (i)–(iii) of Definition 3.1.7 are satisfied with the exception of clause (iia), and we have already noted while performing the T^2-Analysis that clause (i) of Lemma 4.3.1 (Splitting Restraint) is satisfied. Thus all hypotheses of Lemma 4.3.1 (Splitting Restraint) hold, so the conclusion of that lemma implies that shared set action is compatible. By Lemma 4.1.1 (Δ_2 Analysis), all axioms for Δ are declared along a single path through T^1, and whenever an axiom is newly declared at $\rho \subset \Lambda^0$, action for all previously declared axioms is supported at ρ. Furthermore, the compatibility of shared set action implies that the value for an axiom never changes and so that shared functional action is compatible. Hence we can apply Lemma 3.1.6 to conclude that clause (iv) of Definition 2.10.1 holds. The theorem now follows from Theorem 2.11.5 (Modified Framework). ⊣

We note that we can add the requirements of Section 3.3 requiring that A_0 and A_1 be of low degree to our construction, and satisfy these requirements as well. Hence we can strengthen our theorem to:

THEOREM 4.3.4. *Let W be a computably enumerable set. Then there are computably enumerable sets A_0 and A_1 such that if W is not computable, then $W = A_0 \cup A_1$, $A_0 \cap A_1 = \emptyset$, $W \not\leq_T A_0, A_1$, and A_0 and A_1 are low.*

The following corollary follows easily from Theorem 4.3.2.

COROLLARY 4.3.5. *Let $\boldsymbol{a} > \boldsymbol{0}$ be a computably enumerable degree. Then there are computably enumerable degrees $\boldsymbol{a_0}, \boldsymbol{a_1} < \boldsymbol{a}$ such that both $\boldsymbol{a_0} \cup \boldsymbol{a_1} = \boldsymbol{a}$ and $\boldsymbol{a}_0' = \boldsymbol{a}_1' = \boldsymbol{0}'$.*

4.4. Backtracking

The backtracking process addresses the following situation. There will be times, during some constructions, when we are forced to take action for a node $\sigma^m \in T^m$ that does not lie on the current path at η, but was on the current path at some $\xi \subset \eta$. The framework does not allow such action unless we first return σ^m to the current path through T^m as witnessed by one of its derivatives $\kappa \supset \eta$, a process that we call *backtracking*. The precise definition of backtracking will be deferred to Chapter 9, but the definitions and lemmas about backtracking needed to carry out some constructions in the next several chapters are discussed in this section. It is crucial in some constructions, such as permitting constructions, for a node to take advantage of a permission immediately, even when it is not on the current path.

We will introduce a canonical process to obtain κ in Chapter 9, and will call the unique node κ obtained through this process the *backtracking extender* of η for σ^m from ξ. All we need to know right now about κ is that $\kappa \supseteq \eta$ and that $\text{up}^m(\kappa) = \text{up}^m(\xi) = \sigma^m$. It will not always be possible to find such a κ, but the next definition provides a condition that will be shown to be both necessary and sufficient for the existence of κ.

DEFINITION 4.4.1. Fix $\sigma^m \in T^m$ and $\sigma \subseteq \eta \in T^0$ such that σ is an initial derivative of σ^m. Then we say that σ^m is η-accessible with initial derivative σ if no primary η-link properly restrains σ.

If $m \leq 2$, then σ^m will have at most one initial derivative along any $\Lambda^0 \in [T^0]$, so the dependence of the definition of accessibility on the initial derivative will not be necessary, and will be ignored below. In Chapters 4–7, we will only implement backtracking for nodes of T^1 or T^2, so will be able to downplay the dependence of backtracking on a particular initial derivative. The remaining lemmas stated in this section are stated in a form that applies only when $m \leq 2$. The first of these lemmas characterizes the backtracking process, and is a consequence of Lemma 9.3.3 and the fact that initial derivatives of nodes of T^i for $i \leq 2$ are unique.

LEMMA 4.4.2 (Backtracking Lemma). *Fix $m \leq 2$, $\sigma^m \in T^m$ and $\xi \subseteq \eta \in T^0$ such that $\text{up}^m(\xi) = \sigma^m$. Then there is a backtracking extender κ of η for σ^m from ξ iff σ^m is η-accessible. Furthermore, if σ^m is η-accessible, then there is a canonical way to obtain κ.* $\dashv$

When backtracking is implemented, we wish nodes encountered during this process to have the following property.

DEFINITION 4.4.3. Fix $\delta \in T^0$ and $\delta^k \in T^k$. We say that δ^k is *δ-expansionary* if $\text{up}^k((\text{out}^j(\delta^k))^-) = (\delta^k)^-$ for all $j \leq k$. (In other words, for all $j \leq k$, if $\delta^j = \text{out}^j(\delta^k)$ and $\pi^j = (\delta^j)^-$, then π^j is the principal derivative of π^k along $\lambda^j(\delta)$.)

When backtracking is implemented to pass from η to κ, outcomes of the nodes encountered along the way are forced independently of the truth value of the directing sentences of those nodes. The expansionary nodes encountered become *constrained*, as per our next definition.

DEFINITION 4.4.4. Suppose that a construction implements backtracking at η from ξ for $\langle \sigma^m, \sigma \rangle$. Let κ be the backtracking extender of η from ξ for $\langle \sigma^m, \sigma \rangle$, and fix m such that $\sigma^m \in T^m$. We say that δ^k is *constrained* (*for* $\langle \sigma^m, \sigma \rangle$) if $\text{out}^0(\delta^k) \in (\eta, \kappa]$. (Thus the nodes constrained for $\langle \sigma^m, \sigma \rangle$ are those that are δ-expansionary for some $\delta \in (\eta, \kappa]$.) If ν^k is constrained, then it is *finitely constrained* if either k is even and ν^- has Σ outcome along κ or k is odd and ν^- has Π outcome along κ, and is *infinitely constrained* otherwise.

Note that, with notation as in the preceding definition, all nodes in $(\eta, \kappa]$ are constrained.

In order for the backtracking process to be well-behaved, we need to impose restrictions on the situations in which it is allowed to be implemented. When these restrictions are obeyed, we call the backtracking *normal.*

DEFINITION 4.4.5. We say that a construction with true path Λ^0 implements *normal backtracking* if all instances of backtracking within the construction satisfy the following conditions. Suppose that we implement backtracking starting at η for $\langle \sigma^m, \sigma \rangle$ from ξ, that κ is the corresponding backtracking extender for η, and that ρ is the immediate successor of κ along Λ^0. Then:

(i) ρ switches σ^m and $\lambda^m(\rho)$ is ρ-expansionary.
(ii) σ^m has Σ outcome along $\lambda^m(\rho)$.
(iii) Neither the immediate successor of σ along Λ^0 nor ρ is constrained.
(iv) If $\delta \subset \zeta$, either $\lambda^k(\delta)$ is constrained or $(\text{out}^0(\lambda^k(\delta))^-$ is a backtracking extender, and $(\lambda^k(\delta))^-$ is switched by ζ, then ζ is constrained.

As no action is taken for nodes once they become constrained, the faithful execution of a construction requires that constrained nodes are not free at the end of the construction. Our next lemma states that constrained nodes have this property; it is proved as Lemma 9.6.4.

LEMMA 4.4.6 (Constraint Lemma). *Suppose that a given construction implements normal backtracking. Then if ζ^k is constrained,* $\text{out}^0(\zeta^k) \subseteq \xi$ *and ξ is not constrained, then $(\zeta^k)^-$ is not $\lambda^k(\xi)$-free.*

Action will not be supported while a construction implements backtracking. Thus we need the following lemma to ensure that the construction will have infinite support. The lemma is proved as Lemma 9.8.1.

LEMMA 4.4.7. *Fix $\eta^k \subset \Lambda^k \in [T^k]$ such that η^k is Λ^k-free, and suppose that a construction implements only normal backtracking. Then $|\{\rho^k \subset \Lambda^k :$ η^k is ρ^k-free & ρ^k is not constrained$\}| = \infty$.* ⊣

Our last lemma indicates that the iterated backtracking process is finitary and is proved as Lemma 9.8.2.

LEMMA 4.4.8. *Suppose that a construction takes place only on trees T^i for $i \leq 3$. and implements only normal backtracking. Then for any $\eta \subset \Lambda^0$, there is a $\rho \in [\eta, \Lambda^0)$ such that ρ^- is not constrained.* ⊣

4.5. Permitting

The *permitting technique* is a Π_2 technique with a requirement of the form $\Omega(W) = B$, where Player I controls W and Player II must construct B. The computation of B from W requires Player II to define a computable partial functional Ω such that $\Omega(W) = B$, a non-degenerate level 2 requirement. If this is the only non-degenerate level 2 requirement, then we can show that it is satisfied by the construction without placing this requirement on T^2, allowing us to use a Δ_2 construction.

The satisfaction of other requirements will necessitate the entry of numbers x into B after an initial decision is made that $x \notin B$; thus axioms of the form $\Omega(W^s \upharpoonright u; x) = 0$ will already have been defined when the need to place x into B is seen. As Player I controls W, the placement of x into B at stage t requires W to *permit*, i.e., that $W^t \upharpoonright u \neq W^s \upharpoonright u$, else $\Omega(W) \neq B$. But Player II has no control over W, so cannot force an element $< u$ to enter W. The strategy adopted to counteract this problem is to realize that it is not necessary to place the first such x into B; rather, the construction will try to generate infinitely many potential candidates x whose entry into B will ensure the satisfaction of the given requirement, and simultaneously define a partial functional Δ with $\Delta(\emptyset) = W$. Many attempts at defining such a functional will take place, one for each requirement wishing to place a number into B. If one of these requirements is not satisfied, then it will generate infinitely many candidates x, and the failure of W to permit any of these candidates will result in the correctness of the computation of W by $\Delta(\emptyset)$.

The above strategy will succeed only if each requirement wishing to place a number into B can take advantage of every permitting opportunity that materializes. In prior constructions, it was sufficient to offer a requirement infinitely many opportunities for satisfaction; but the success of permitting necessitates that the requirement be offered cofinitely many opportunities for satisfaction. Derivatives of a node $\eta^2 \in T^2$ can act only when they lie on the current path, and the permissions for such a node may occur when it does not lie on the current path. The solution is to suspend action for the construction until we backtrack, thereby returning the node for which we want to act to the current path. Action must be suspended during backtracking, as it could destroy the validity of the directing sentence causing us to want to backtrack. It will turn out that the nodes for which we fail to take action are not free at the end of the construction, so we can still argue that the construction is faithfully executed.

We now prove a *Permitting Lemma* that allows us to conclude that the permitting strategy will succeed.

LEMMA 4.5.1 (Permitting Lemma). *Suppose that a construction with true path Λ^0 has the following properties for all $\eta \subset \nu \subseteq \rho \subset \Lambda^0$ such that $\nu^- = \eta$:*

(*i*) *$\langle\, \rangle$ is not constrained and $B^{\mathrm{wt}(\langle\,\rangle)} = \emptyset$.*
(*ii*) *If $x \in B^{\mathrm{wt}(\nu)} - B^{\mathrm{wt}(\eta)}$ and σ is the longest node $\subset \eta$ that is not constrained, then $W^{\mathrm{wt}(\eta)} \upharpoonright x+1 \neq W^{\mathrm{wt}(\sigma)} \upharpoonright x+1$.*
(*iii*) *There are infinitely many $\sigma \subset \Lambda^0$ that are not constrained.*

Then $B \leq_T W$.

PROOF. Fix x. By (iii), we may fix the shortest non-constrained $\eta \subset \Lambda^0$ such that $W^{\mathrm{wt}(\eta)} \restriction x+1 = W \restriction x+1$, and $\nu \subset \Lambda^0$ such that $\nu^- = \eta$. Note that η and ν are effectively determined by W. By (i), there is a longest non-constrained $\sigma \subset \eta$, so by (ii), $x \in B$ iff $x \in B^{\mathrm{wt}(\nu)}$. ⊣

The Permitting Lemma is now used to pull the Friedberg–Mučnik Theorem down below any non-computable, computably enumerable degree, and so to show that no computably enumerable degree is minimal. The permitting technique was introduced by Friedberg [4] to obtain this result.

THEOREM 4.5.2. *Let W be a computably enumerable set. Then either W is computable, or there are computably enumerable sets $A_0, A_1 \leq_T W$ whose degrees are incomparable.*

The next corollary follows easily from Theorem 4.5.2.

COROLLARY 4.5.3. *For every computably enumerable degree $\mathbf{d} > \mathbf{0}$, there are computably enumerable degrees $\mathbf{a_0}, \mathbf{a_1} < \mathbf{d}$ such that $\mathbf{a_0} \mid \mathbf{a_1}$.*

PROOF OF THEOREM 4.5.2. We will define computable partial functionals Ω_0 and Ω_1 such that $A_0 = \Omega_0(W)$ and $A_1 = \Omega_1(W)$. In addition, for each computable partial functional Φ and each $i \leq 1$, we establish the requirement

$$P_{\Phi,i} : \Phi(A_i) = A_{1-i} \rightarrow \exists\Delta(\Delta(\emptyset) = W).$$

The requirements $P_{\Phi,i}$ are organized into an effective list $\{R_i : i \in \mathbb{N}\}$.

4.5.1. The Basic $P_{\Phi,i}$-Module. This module has a shortest node α with $\mathrm{lev}(\alpha) = 1$. α will have a non-terminal, level 2, immediate successor α_0 along which it is activated and has Σ outcome, and a terminal immediate successor α_1 along which it is validated and has Π outcome. The non-terminal node α_0 will have a terminal immediate successor α_{00} (α_{01}, resp.) along which it is activated (validated, resp.) and has Σ outcome (Π outcome, resp.). It is easy to see that clauses (i)–(iii) of Definition 2.3.1 are satisfied.

The directing sentence for α checks to see if there is a permitted derivative η^1 of α_0. As it will be doing its checking along Λ^0, the parameters already determined for α_0 and its derivatives can be used in this check. Thus it is more instructive to analyze the situation for α_0 first.

The directing sentence for α_0 checks to see if $\Phi(A_i; \mathrm{wt}(\eta^1)) \downarrow = 0$ for all $\eta^1 \in \widetilde{V}_\alpha$. Activated action for α_0 specifies that $\mathrm{wt}(\eta^1) \notin A_{1-i}$, and validated action for α_0 restrains A_i and defines $\Delta(\emptyset) = W$. The witnesses u_{η^1} and $\eta^0_{\eta^1}$ for η^1 can be used in the action sentences for α_0 and also in the sentences for α as long as they have been determined beforehand.

Directing sentence S_{α_0}:

$$\forall\eta^1 \in \widetilde{V}_{\alpha_0} \exists u \exists\eta^0 \in \widetilde{V}_{\eta^1}(\Phi(A_i \restriction u; \mathrm{wt}(\eta^1))[\mathrm{wt}(\eta^0)] = 0).$$

Activated action for α_0 at $\rho \supseteq \alpha_{00}$: $\forall\eta^1 \in \widetilde{V}_{\alpha_0} \forall t(\mathrm{wt}(\eta^1) \notin A^t_{1-i})$.

Validated action for α_0 at $\rho \supseteq \alpha_{01}$:

$$\forall \eta^1 \in \widetilde{V}_{\alpha_0} \exists r \forall t \geq r \forall p \geq \mathrm{wt}(\eta^0_{\eta^1}) \forall y \leq \mathrm{wt}(\eta^1)$$

$$(\Delta^r_{\alpha_0}(\emptyset; y) \downarrow = W^p(y) \;\&\; A^t_i \upharpoonright u_{\eta^1} = A_i^{\mathrm{wt}(\eta^0_{\eta^1})} \upharpoonright u_{\eta^1}).$$

Returning to α, we first present a definition. We say that $\xi^1 \in T^1$ *requires attention for* $\eta^1 \in T^1$ at $\sigma \in T^0$ if ξ^1 is an α-node, $(\mathrm{up}(\eta^1))^- = \mathrm{up}(\xi^1)$, η^1 has Σ outcome along $\lambda(\sigma)$, and $W^{\mathrm{wt}(\eta^0_{\eta^1})} \upharpoonright \mathrm{wt}(\eta^1) + 1 \neq W^{\mathrm{wt}(\sigma)} \upharpoonright \mathrm{wt}(\eta^1) + 1$.

The directing sentence for α searches for permission from W for one of the α_0-nodes, $\eta^1 \in T^1$, from its module. Activated action specifies that $\mathrm{wt}(\eta^1) \notin A_{1-i}$, and validated action specifies that $\mathrm{wt}(\eta^1) \in A_{1-i}$. As α-nodes are level 1 nodes, we will interpret the parameter α in its directing sentence on T^1. There is no activated action for α. The witness η^1 for α can be used in its validated action sentence.

Directing sentence S_α: $\exists \eta^1 \exists \sigma \in \widetilde{V}_\alpha(\alpha$ requires attention for η^1 at $\sigma)$.

Validated action for α at $\rho \supseteq \alpha_1$:

$$\exists r \forall t \geq \mathrm{wt}(\eta^0_{\eta^1})(\mathrm{wt}(\eta^1) \in A^r_{1-i} \;\&$$

$$\Phi(A_i \upharpoonright u_{\eta^1}; \mathrm{wt}(\eta^1))[\mathrm{wt}(\eta^0_{\eta^1})] = 0 \;\&\; A^t_i \upharpoonright u_{\eta^1} = A_i^{\mathrm{wt}(\eta^0_{\eta^1})} \upharpoonright u_{\eta^1}).$$

We now see that $P_{\Phi,i}$ is satisfied when $\alpha \subset \Lambda^2$ and sentences are faithfully interpreted. For if $\alpha_1 \subset \Lambda^2$, then there is an $\eta^1 \in \widetilde{V}_{\alpha_0}$ such that $\Phi(A_i; \mathrm{wt}(\eta^1)) \downarrow = 0 \neq 1 = A_{1-i}(\mathrm{wt}(\eta^1))$; if $\alpha_{00} \subset \Lambda^2$, then either there is an $\eta^1 \in \widetilde{V}_{\alpha_1}$ such that $\Phi(A_i; \mathrm{wt}(\eta^1)) \downarrow \neq 0 = A_{1-i}(\mathrm{wt}(\eta^1))$, or $\Phi(A_i)$ is not total; and if $\alpha_{01} \subset \Lambda^2$, then by Lemmas 3.1.8(iv) and 4.4.6, W is computable.

4.5.2. T^2-Analysis. Let $\eta^2 \in T^2$ be given such that requirements have been assigned to all $\xi^2 \subset \eta^2$ but no requirement has yet been assigned to η^2. The standard initial assignment process is followed at η^2. Clause (i) of Definition 2.7.1 follows from Lemma 2.7.4 (Standard Initial Assignment).

The standard initial sentence specification for $\eta^2 \in T^2$ is followed, with action supported at $\{\rho^2 \supset \eta^2 : \eta^2$ is ρ^2-free$\}$. Clause (i) of Definition 2.10.1 will follow from the analysis for the basic modules and Lemma 4.5.1 (Permitting), once we verify the hypotheses of the Permitting Lemma. It will then suffice to show that all requirements are satisfied.

4.5.3. T^1-Analysis. The standard derived assignment is followed. Clauses (i)–(iii) of Definition 2.7.7 follow from Lemma 2.12.2 (Standard Derived Assignment).

Sentence decomposition for nodes $\eta^1 \in T^1$ bounds η^1 in the sentences with the node η^1 currently being visited. Monotonicity of sentence decomposition follows from (2.8.1). Action for an α-node or an α_0-node $\eta^1 \in T^1$ is supported at $\{\rho^1 \supset \eta^1 : \eta^1$ is ρ^1-free & ρ^1 is not ρ^1-constrained$\}$. By Lemma 4.4.7, the construction has infinite support on T^1.

4.5.4. T^0-Analysis. The standard derived assignment is followed. Clauses (i)–(iii) of Definition 2.7.7 follow from Lemma 2.12.2 (Standard Derived Assignment).

Fix $\eta \subset \nu \subseteq \rho \subset \Lambda^0$ such that $\nu^- = \eta$. Sentence decomposition is as follows. $\eta^0 \subset \eta$, $u \leq \text{wt}(\lambda(\eta))$, $r = \text{wt}(\nu)$, $t \leq \text{wt}(\rho)$, $\sigma = \eta$, $\eta^1 \subseteq \lambda(\eta)$ and $p = \text{wt}(\delta)$ where δ is the longest node $\subset \rho$ that is not ρ-constrained. Monotonicity of sentence decomposition follows from (2.8.1) and (2.8.4). Action for η is supported at $\{\rho \supset \eta : \eta \text{ is } \rho\text{-free} \ \& \ \rho \text{ is not } \rho\text{-constrained}\}$. By Lemma 4.4.7, the construction has infinite support on T^0.

4.5.5. The Construction. Suppose that we are at the node $\eta \subset \Lambda^0$. If there are ξ^1 and η^1 such that ξ^1 requires attention for η^1 at η, fix the shortest such ξ^1 and fix the initial derivative ξ of ξ^1 along η. If $\xi^1 \neq \text{up}(\eta)$, let κ be the backtracking extender of η for ξ^1 from ξ (Lemma 4.4.2 (Backtracking) ensures the existence of κ); we specify that $\kappa \subset \Lambda^0$. Otherwise, or if $\xi^1 = \text{up}(\eta)$ in the preceding sentence, the construction acts in accordance with clause (iva) of Definition 2.10.1 at η.

4.5.6. The Verification. Fix the true path $\Lambda^0 \in [T^0]$ determined by the construction, and let $\Lambda^{k+1} = \lambda(\Lambda^k)$ for $k \leq 1$. It is easily seen that the construction admits consistent action, and that backtracking is normal. By Lemma 4.4.6, the construction is faithfully executed, and so by Lemma 3.1.8(ii) (Σ_1 Analysis), support propagates from T^0 to T^1 and from T^1 to T^2. We define computable partial functionals Ω_0 and Ω_1 at nodes ρ such that ρ^- is not constrained, setting

$$\Omega_0(W;x)[\text{wt}(\rho)] = A_0^{\text{wt}(\rho)}(x) \text{ and } \Omega_1(W;x)[\text{wt}(\rho)] = A_1^{\text{wt}(\rho)}(x)$$

for all $x \leq \text{wt}(\text{up}(\eta))$ for some $\eta \subset \rho$. By Lemma 4.4.8, both $\Omega_0(A)$ and $\Omega_1(A)$ will be total. The satisfaction of the hypotheses of Lemma 4.5.1 (Permitting) is an easy consequence of backtracking, so as noted above, clause (i) of Definition 2.10.1 is satisfied.

If $\eta^2 \subset \Lambda^2$ is an α_0-node, then new axioms for $\Delta = \Delta_{\alpha_0}$ are declared only when $\Delta_{\alpha_0}(\emptyset)$ and W are compatible (else we would implement backtracking which would force $\eta^2 \not\subset \Lambda^2$), so shared functional axioms for Δ_{α_0} are compatible. We note that if we were to replace the restraint condition

$$\forall t \geq \text{wt}(\eta^0_{\eta^1})(A_i^t \upharpoonright u_{\eta^1} = A_i^{\text{wt}(\eta^0_{\eta^1})} \upharpoonright u_{\eta^1})$$

at an α-node ξ with

$$\forall t \geq \text{wt}(\xi)(A_i^t \upharpoonright u_{\eta^1} = A_i^{\text{wt}(\xi)} \upharpoonright u_{\eta^1}),$$

then clauses (ii) and (iii) of Definition 3.1.7 would hold, so we would be able to apply Lemma 3.1.9(ii) (Restraint) to conclude that shared set action is compatible. But by Lemma 3.1.9(i) (Restraint), $A_i^{\text{wt}(\xi)} \upharpoonright u_{\eta^1} = A_i^{\text{wt}(\eta^0_{\eta^1})} \upharpoonright u_{\eta^1}$; hence we can conclude that shared set action is compatible. Lemma 3.1.6

now allows us to conclude that clause (iv) of Definition 2.10.1 holds, so the theorem follows from Theorem 2.11.5 (Modified Framework). ⊣

4.5.7. Further Remarks. It is easy to see that the lowness requirements of Section 3.3 can be combined with this construction. It can thus be shown that:

THEOREM 4.5.4. *Let W be a computably enumerable set. Then either W is computable, or there are low computably enumerable sets $A_0, A_1 \leq_T W$ whose degrees are incomparable.*

We note that the permitting techniques introduced above can be used to construct a properly d-c.e. degree below any non-zero computably enumerable degree. A proof of this fact is left to the reader. In general, permitting can be combined with a construction that has the ability to utilize the permission almost immediately (as measured by a computable function).

4.6. Summary

The introduction of requirements at the Δ_2 level presented us with a new situation in which we had level 2 nodes responsible for the declaration of axioms for functionals $\Delta(\emptyset)$. In Sections 4.2 and 4.3, the axioms to be declared required us to copy values produced by another functional whose oracle A was a set controlled by Player II. The imposition of restraint on A allowed us to keep the values specified for Δ from changing, and so to show that shared axioms for Δ were compatible.

In Section 4.6, however, the values for $\Delta(\emptyset)$ were produced by a computably enumerable set W controlled by Player I. In order to avoid instructions to declare incompatible axioms, the instructions for Δ to stop declaring axioms had to be implemented immediately upon finding a change in the value produced by W. As this change might be found when the node under consideration was not responsible for declaring axioms for W, we needed to implement backtracking to reach such a node, and suspend all action for the construction in the interim. This will have brought us to a later stage, at which we might need to backtrack for a different node of T^1. However, we can restrict the possibility of iterating backtracking to successively shorter nodes of T^1, so the process will terminate.

In the next section, we will have level 2 requirements whose action instructs Player II to declare axioms for functionals whose oracles are controlled by Player I, or to copy values of functionals controlled by Player I. New techniques will be required to resolve conflicts between shared functional axioms in that situation.

Chapter 5

Π_2 CONSTRUCTIONS

We now turn our attention to theorems whose proofs use Π_2 constructions. We begin, in Section 5.1, by stating lemmas that isolate combinatorial properties of Π_2 constructions. We then implement Π_2 constructions to prove the existence of a high computably enumerable degree in Section 5.2, the Sacks Jump Inversion Theorem in Section 5.3, the Minimal Pair Theorem in Section 5.4 and an embedding of the pentagon into the computably enumerable degrees in Section 5.5.

5.1. Π_2 Constructions

Requirements such as the thickness requirements discussed in Chapter 2 act to construct functionals that are total on given oracles, so require the declaration of infinitely many axioms. Such requirements cannot be handled by level 1 constructions, as a node of T^1 will generally have the responsibility to declare only finitely many axioms for such functionals. This will introduce a coordination problem, as nodes having the responsibility to declare axioms on the same argument will appear on incomparable paths through T^1.

Some requirements of the form $\Phi(A) = W$ can be handled by level 2 constructions. These constructions capture what are traditionally called *infinite injury constructions.* For those familiar with infinite injury constructions, we describe how these constructions are simulated by level 2 constructions.

A typical infinite injury construction will assign requirements to nodes of a tree T^2, that can always be taken to be a binary tree. The nodes of the tree are prioritized, using the lexicographical ordering of these nodes. A computation of the current path through T^2 is made at each stage of the construction, and action is carried out for nodes that lie along the current path. At the end of the construction, a true path $\Lambda^2 \in T^2$ is obtained. Nodes to the right of Λ^2 have lower priority than any nodes along Λ^2 and are initialized each time a node to their left is visited; this corresponds, in a level 2 argument, to the fact that the initial derivative, on T^1, of a node to the right of Λ^2 will have weight larger than the use of any axiom declared at that point of the construction for a derivative of a node along Λ^2. A node to the left of Λ^2 will act only finitely often in an infinite injury construction; this corresponds to the fact that there are only finitely many $\eta \subset \Lambda^0$ (the true path through T^0 in the iterated trees of strategies approach) at which derivatives of a fixed node to the left of Λ^2 are free. Along Λ^2, longer nodes can act only when their guesses at the outcomes of shorter nodes seem to be correct; this is ensured, in the iterated trees of strategies approach, by Definition 2.4.1 and (2.6.4).

The situation with shared functional axioms becomes more complex on T^2, as clause (iv) of Definition 3.1.7 may no longer hold. Thus a new restraint lemma will be needed, and will be proved as Lemma 8.8.4.

LEMMA 5.1.1 (Π_2 Restraint Lemma). *Suppose that for a given faithfully executed construction, action for η is supported at ρ only when η is ρ-free, sentence decomposition is monotonic, the construction admits consistent action, and clauses (i)–(iii) of Definition* 3.1.7, *with equality in (ia), are satisfied. Assume also that the following condition is satisfied for the functional Δ with oracle A.*

(*i*) *Suppose that $\langle \eta_0^2, \rho_0^2 \rangle$ and $\langle \eta_1^2, \rho_1^2 \rangle$ share the functional axiom $\Delta(A; x)$, and let $\kappa^2 = \eta_0^2 \wedge \eta_1^2$. Then:*
  (*a*) *If $\eta_0^2 \neq \eta_1^2$, then $\eta_0^2 | \eta_1^2$.*
  (*b*) *If $\mathrm{lev}(\kappa^2) \geq 2$, and $\kappa^1 \in T^1$ is a derivative of κ^2, then validated action for κ^1 requires that $\mathrm{wt}(\kappa^1)$ be placed into A, and no other requirement on T^1 has action placing $\mathrm{wt}(\kappa^1) \in A$.*
  (*c*) *For $i \leq 1$, if action for η_i^1 is supported at ρ_i^1 and $\mathrm{up}(\eta_i^1) = \eta_i^2$, then $\mathrm{wt}(\eta_i^1) \leq x < \mathrm{wt}(\rho_i^1)$.*

Then shared functional action for $\Delta(A)$ is compatible.

In constructions where we are asked to define axioms from an oracle W over which we have no control, we will not be able to correct axioms. One method to circumvent this problem, applied in Section 5.3, is to ensure that distinct nodes of T^2 generate incompatible guesses at the oracle. We will have to differentiate between nodes of level 2 and those of level < 2; for convenience, we assume that nodes on T^2 of level 2 have even length, and nodes of level < 2 have odd length. (This assumption can be easily modified to cover constructions where the requirements (or basic modules) are not as uniformly distributed on T^2.) We first define a 0–1-valued string χ_{η^2} that codes whether the nodes $\subset \eta^2$ of level 2 have Π or Σ outcome along η^2, and show that the definition can distinguish outcomes as described. We then generate another string $\gamma_{\mathrm{out}(\sigma^2)}$ that allows us to distinguish between like outcomes when we pass to T^1. Lemma 5.1.5 formalizes this distinction.

DEFINITION 5.1.2. Fix $\eta^2 \in T^2$ such that $|\eta^2| = 2e+2$. Define the string χ_{η^2} of length $e+1$ by $\chi_{\eta^2}(x) = 0$ ($= 1$, resp.) iff $\eta^2 \restriction 2x$ is activated (validated, resp.) along η^2. η^2 will generate the guess that $\chi_{\eta^2} \subset W$.

It is immediate from Definition 5.1.2 that:

LEMMA 5.1.3. *Suppose that* $\eta^2, \xi^2 \in T^2$ *and* $|\eta^2| = |\xi^2| = 2e + 2$. *Then the following are equivalent*:

(*i*) χ_{η^2} *and* χ_{ξ^2} *are compatible.*

(*ii*) $\chi_{\eta^2} = \chi_{\xi^2}$.

(*iii*) *For all* $x < e + 1$, $\eta^2 \restriction 2x$ *is activated along* η^2 *iff* $\xi^2 \restriction 2x$ *is activated along* ξ^2. ⊣

Under suitable hypotheses, incompatible paths through T^2 of the same even length will generate incompatible guesses at oracles $W \oplus V$, where V is a path through T^1 determined by the path through T^2. Guesses at W will be of the form $\chi_{\eta^2} \subset W$ for $\eta^2 \in T^2$. We will also define a string $\gamma_{\text{out}(\eta^2)}$, and will require that $\gamma_{\text{out}(\eta^2)} \subset V$. We use $\text{out}(\eta^2)$ as the subscript for γ, as we wish the string on T^1 to depend only on η^2, so can use only as much of the true path through T^1 as is ensured by having η^2 lie on the true path through T^2.

DEFINITION 5.1.4. Fix $\nu^1 \in T^1$. Define the string γ_{ν^1} of length $|\nu^1|$ by $\gamma_{\nu^1}(x) = 0$ ($= 1$, resp.) iff $\nu^1 \restriction x$ is activated (validated, resp.) along ν^1. ν^1 will generate the guess that $\gamma_{\nu^1} \subset V$.

The following lemma will be proved as Lemma 8.9.3.

LEMMA 5.1.5 (Incompatible Oracle Lemma). *Let* $\xi \subset \eta \subset \Lambda^0 \in [T^0]$ *and* $\eta^2 \neq \xi^2 \in T^2$ *be given such that* $\xi = \text{out}^0(\xi^2)$, $\eta = \text{out}^0(\eta^2)$, *and* $|\eta^2| = |\xi^2| = 2e + 2$. *Suppose that for all* $\rho^2 \in T^2$ *such that* $|\rho^2|$ *is odd*, $\text{lev}(\rho^2) < 2$. *Then either* χ_{η^2} *and* χ_{ξ^2} *are incompatible, or* $\gamma_{\text{out}(\eta^2)}$ *and* $\gamma_{\text{out}(\xi^2)}$ *are incompatible.* ⊣

5.2. A High Computably Enumerable Degree

The existence of a high computably enumerable degree, proved by Sacks [22], has a construction that begins on T^2 and implements infinitary action to place elements into sets. This forces action for a node of T^1 to take place at many of its successors along a path, rather than just at the immediate successor. Recall that K is a computably enumerable set of degree $\boldsymbol{0}'$. To show that H is high, we compute K' from H via a limit approximation. This computation must be correct no matter which path through T^2 is followed, so we will have nodes on T^0 derived from incomparable nodes of T^2 declaring axioms for the same functional. (We call a requirement with this property a *global requirement*.) In order for the construction to succeed, we must coordinate the declaration of axioms by different nodes, else when we switch paths, we may already have declared an axiom that conflicts with the axiom we now wish to declare.

The solution is to implement the strategy employed in Theorem 3.3.1. We require that when a derivative, η^1, of a level 2 node η^2 is validated on T^1, then $\text{wt}(\eta^1)$ is placed into all sets of high degree that are being constructed (this will be possible even if we are constructing incomparable high degrees). Such action will automatically correct axioms when paths are switched on T^1, unless the path is being switched by a level 1 node, in which case η^2 will not be on the true path through T^2. The only other way to switch paths is to pass to paths on T^2 that have not previously been visited, and correction will not be necessary in this case.

THEOREM 5.2.1 (High Computably Enumerable Degree Theorem). *There is a computably enumerable set $H <_T K$ of high degree.*

PROOF. We will construct a computably enumerable set H of high degree. In order to ensure that $K \not\leq_T H$, we construct a second computably enumerable set D and satisfy the following level 1 requirement for all computable partial functionals Φ:

$$P_\Phi : \Phi(H) \neq D.$$

The predicate $x \in K'$ is Σ_2, so there is a computable predicate R such that

$$x \in K' \Leftrightarrow \exists w \forall v \neg R(w, v, x).$$

In order to show that H is high, we define a computable partial functional Δ that satisfies the following level 2 requirement for each $x \in \mathbb{N}$:

$$Q_x : \lim_u \Delta(H; x, u) = \begin{cases} 1 & \text{if } x \in K', \\ 0 & \text{if } x \notin K'. \end{cases}$$

(Recall that we do not require $\Delta(H)$ to be total when we apply the Limit Lemma.) Let $\{R_i : i \in \mathbb{N}\}$ be an effective list of all requirements.

5.2.1. The Basic P_Φ-Module. This is essentially the incomparability module of Theorem 3.2.1. It is assigned to α-nodes of T^2 that have level 1, and the parameter $\text{wt}(\alpha)$ is replaced by the weight of a derivative α^1 of α on T^1, which is quantified over existentially in the directing sentence.

Directing sentence S_α:

$$\exists \alpha^1 \in V_\alpha \exists u \exists \eta \in V_{\alpha^1} (\Phi(H \restriction u; \text{wt}(\alpha^1))[\text{wt}(\eta)] \downarrow = 0).$$

Activated action for α at $\rho \supseteq \alpha_0$: $\forall t (x \notin D^t)$.

Validated action for α at $\rho \supseteq \alpha_1$:

$$\exists r \forall t \geq \text{wt}(\eta)(x \in D^r \ \& \ H^t \restriction u = H^{\text{wt}(\eta)} \restriction u).$$

The parameter η for validated action is supplied by the directing sentence. Clauses (i)–(iii) of Definition 2.3.1 follow easily. If S_α is true and validated action is implemented at extensions of α, then there is an x such that $\Phi(H; x) \downarrow = 0 \neq 1 = D(x)$; and if S_α is false and activated action is implemented at extensions of α, then there is an x such that either $\Phi(H; x) \uparrow$, or $\Phi(H; x) \downarrow \neq 0 = D(x)$. In either case, P_Φ will be satisfied.

5.2.2. The Basic Q_x-Module. This module consists of a single non-terminal level 2 node β, with terminal extensions β_0, along which β is activated and has Σ outcome, and β_1, along which β is validated and has Π outcome. Clauses (i)–(iii) of Definition 2.3.1 follow easily. We define the parameter $p(\beta)$ to be $\mathrm{wt}(\mathrm{init}^1(\beta, \mathrm{out}(\beta_0)))$ below.

Directing sentence S_β: $\forall w \exists v (R(w, v, x))$.

Activated action for β at $\rho \supseteq \beta_0$:

$$\forall y \geq p(\beta) \exists u \exists r \forall t > r (\Delta(H^r \upharpoonright u; x, y) = 1 \,\&\, H^t \upharpoonright u = H^r \upharpoonright u).$$

Validated action for β at $\rho \supseteq \beta_1$:

$$\forall y \geq p(\beta) \exists u \exists r \forall t > r (\Delta(H^r \upharpoonright u; x, y) = 0 \,\&\, H^t \upharpoonright u = H^r \upharpoonright u).$$

If S_β is true and validated action is implemented at extensions of β, then $x \notin K'$ and, by the Limit Lemma, $\lim_u \Delta(H; x, u) = 0$; and if S_β is false and activated action is implemented at extensions of β, then $x \in K'$ and, by the Limit Lemma, $\lim_u \Delta(H; x, u) = 1$. In either case, Q_x will be satisfied.

5.2.3. T^2-Analysis. The standard initial assignment process is followed. Clause (i) of Definition 2.7.2 follows easily from Lemma 2.7.4 (Standard Initial Assignment). Hence it suffices to show that all requirements are satisfied.

The standard initial sentence specification for $\eta^2 \in T^2$ is followed, with action for η^2 supported at $\{\rho^2 \supset \eta^2 : \eta^2 \text{ is } \rho^2\text{-free}\}$. Clause (i) of Definition 2.10.1 follows from the analysis for basic modules.

5.2.4. T^1-Analysis. The standard derived assignment is followed. Clauses (i)–(iii) of Definition 2.7.7 follow from Lemma 2.12.2 (Standard Derived Assignment).

The variable α^1 in the directing sentence S_α is assigned the witness $\mathrm{wt}(\eta^1)$ when the sentence is assigned to $\eta^1 \in T^1$. We add a condition to the action sentences for β nodes η^1, specifying sentence decomposition as follows:

Directing sentence S_{η^1}: $\forall w \leq \mathrm{wt}(\eta^1) \exists v (R(w, v, x))$.

Activated action for η^1 at $\rho \supset \eta^1$:

$$\forall y \in [p(\mathrm{up}(\eta^1)), \mathrm{wt}(\eta^1)] \exists u \exists r \forall t > r$$
$$(\Delta(H^r \upharpoonright u; x, y) = 1 \,\&\, H^t \upharpoonright u = H^r \upharpoonright u \,\&\, \mathrm{wt}(\eta^1) \notin H^t).$$

Validated action for η^1 at $\rho \supset \eta^1$:

$$\forall y \in [p(\mathrm{up}(\eta^1)), \mathrm{wt}(\eta^1)] \exists u \exists r \forall t > r$$
$$(\Delta(H^r \upharpoonright u; x, y) = 0 \,\&\, H^t \upharpoonright u = H^r \upharpoonright u \,\&\, \mathrm{wt}(\eta^1) \in H^r).$$

(The validated action clause $\mathrm{wt}(\eta^1) \in H^r$, may seem 'capricious', (terminology introduced by Harrington), as it is irrelevant to the above proof of the satisfaction of the requirement. However, it will be crucial in the application of Lemma 3.1.9 (Restraint).)

Monotonicity of sentence decomposition follows from (2.8.1). Action for $\eta^1 \in T^1$ is supported at $\{\rho^1 \supset \eta^1 : \eta^1 \text{ is } \rho^1\text{-free}\}$. By Lemma 3.1.8(i,ii) ($\Sigma_1$ Analysis), the construction has infinite support on T^1 and support propagates from T^1 to T^2.

5.2.5. T^0-Analysis. The standard derived assignment is followed. Clauses (i)–(iii) of Definition 2.7.7 follow from Lemma 2.12.2 (Standard Derived Assignment).

Fix $\eta = v^- \subset v \subseteq \rho \in T^0$, and let $\eta^1 = \text{up}(\eta)$. Sentence decomposition for α-nodes is as in Theorem 3.2.1 (Friedberg–Mučnik). Sentence decomposition for β-nodes η is as follows: $v \leq \text{wt}(\eta)$, $u \leq \text{wt}(\lambda(\eta))$, $r = \text{wt}(v)$, and $t \leq \text{wt}(\eta)$. Monotonicity of sentence decomposition follows from (2.8.1) and (2.8.4). Action for η is supported at $\{\rho \supset \eta : \eta \text{ is } \rho\text{-free}\}$. By Lemma 3.1.8(i,ii) ($\Sigma_1$ Analysis), the construction has infinite support on T^0 and support propagates from T^0 to T^1.

5.2.6. The Construction. The construction is executed in accordance with clause (iva) of Definition 2.10.1.

5.2.7. The Verification. Fix the true path $\Lambda^0 \in [T^0]$ determined by the construction, and for $i \leq 2$, let $\Lambda^i = \lambda^i(\Lambda^0)$. It is easily seen that the construction admits consistent action. Clauses (i)–(iii) of Definition 3.1.7 are routine to verify, so by Lemma 3.1.9 (Restraint), shared set action is compatible. Furthermore, hypothesis (i) of Lemma 5.1.1 is satisfied, so by Lemma 5.1.1 (Π_2-Restraint), shared functional action is compatible. We now see that the hypotheses of Lemma 3.1.6 hold. Thus by Lemma 3.1.6, clause (iv) of Definition 2.10.1 holds, so the theorem follows from Theorem 2.11.5 (Modified Framework). ⊣

5.2.8. Further Remarks. In Section 3.1, we discussed the components of requirements used to prove theorems about the poset $\mathcal{R}$. When the jump operator is introduced, we must expand the types of requirements considered. One way to do so, as was used in the proof of the preceding theorem, is to deal with functionals having larger numbers of variables, and take the limit operation. An alternate way would have been to define a partial functional Δ, with requirements that $\Delta(H; x) \downarrow$ iff $x \in K'$. The latter is hard to iterate, however, when the jump operator is applied several times. In such cases, we can either use iterated limits, as was done in [13], or build sets of successively higher levels that are computably enumerable in the sets of lower levels. This seems to be the easier way, and will be adopted later in this monograph.

5.3. The Jump Inversion Theorem

One of the first theorems proved by Sacks [22] using the infinite injury priority method was the Jump Inversion Theorem. This theorem provides a necessary and sufficient condition for a degree to be the jump of a computably enumerable degree, namely, that the degree be $\geq \mathbf{0}'$ and computably enumerable in $\mathbf{0}'$. The necessity of this condition was known – the sufficiency of the condition is the content of the Jump Inversion Theorem. In fact, Sacks proved a more general version of this theorem from which the existence of a high computably enumerable degree (Theorem 5.2.1) follows. We present a proof of the basic Jump Inversion Theorem using the Framework.

Given a Σ_2 set $C \geq_T \emptyset'$, we adopt the same strategy as in Theorem 5.2.1 (High Computably Enumerable Degree) to ensure that $C \leq_T A'$. We need a new strategy, however, to ensure that $A' \leq_T C$. We will define a set $D \leq_T \emptyset'$ (D will be computed from the true path Λ^1 through T^1) and a computable partial functional Θ such that for all e, $\Theta(C, D; e) \downarrow = A'(e)$. As we have no control over C, we will not be able to correct axioms. Thus we try to ensure that distinct nodes of T^2 of length $2e + 2$ generate incompatible guesses at the C oracle, using the strings of Definition 5.1.2, with the final path Λ^2 through T^2 generating the correct guess. The guesses at C, however, do not allow sufficient flexibility for incompatible paths through T^2 to declare correct axioms from incompatible oracles. However, if $\sigma^2 \neq \tau^2$, then $\text{out}(\sigma^2) \neq \text{out}(\tau^2)$. We use $\text{out}(\sigma^2)$ to generate a guess at a set $D \leq_T \emptyset'$ in such a way that if $\chi_{\sigma^2} \neq \chi_{\tau^2}$, then the guesses for initial segments of D generated by σ^2 and τ^2 are incompatible. These guesses are generated by Definition 5.1.4. Lemma 5.1.5 (Incompatible Oracle) will then be used to show that we have succeeded in having $\Theta(C, D; e)$ compute $A'(e)$.

Lemma 5.1.5 (Incompatible Oracle) will be used as follows. Once we have the true path Λ^0 for the construction, with $\Lambda^2 = \lambda^2(\Lambda^0)$, we will show that if $\sigma^2 \subset \Lambda^2$, then χ_{σ^2} is a substring of C. Furthermore, by Remark 2.4.2(i), $\gamma_{\text{out}(\sigma^2)}$ is uniformly computable from $\Lambda^1 = \lambda(\Lambda^0)$. As $\Lambda^1 \leq_T \emptyset'$, Lemma 5.1.5 (Incompatible Oracle) implies that incompatible strings on T^2 that are visited during the construction generate axioms from incompatible oracles. Thus we can ensure that Θ generates axioms that are correct.

THEOREM 5.3.1. *Let $\mathbf{c} \geq \mathbf{0}'$ be computably enumerable in $\mathbf{0}'$. Then there is a computably enumerable degree $\mathbf{a}$ such that $\mathbf{a}' = \mathbf{c}$.*

PROOF. Since $\mathbf{c}$ is computably enumerable in $\mathbf{0}'$, there is a Σ_2 set C of degree $\mathbf{c}$. Thus we can fix a computable predicate R such that for all $e \in \mathbb{N}$,

$$e \in C \Leftrightarrow \exists u \forall v (\neg R(u, v, e)). \tag{5.3.1}$$

We construct a computably enumerable set A such that $A' \equiv_T C$, another computably enumerable set D, and computable partial functionals Δ and Θ

that satisfy the following requirements for all $e \in \mathbb{N}$:

$$Q_e : \lim_u \Delta(A; u, e) = C(e).$$

$$N_e : A'(e) = \Theta(C, D; e).$$

We establish an effective list $\{R_e : e \in \mathbb{N}\}$ of all such requirements, with $R_{2e} = Q_e$, and $R_{2e+1} = N_e$. By the Limit Lemma (recall that we do not require $\Delta(A)$ to be total when we apply the Limit Lemma), if Q_e is satisfied for all $e \in \mathbb{N}$, then $C \leq_T A'$. As $C \geq 0'$, $C \oplus D \equiv_T C$, so we can satisfy $A' \leq_T C$ by showing instead that $A' \leq_T C \oplus D$. The latter follows from the satisfaction of N_e for all $e \in \mathbb{N}$.

5.3.1. The Basic Q_e-Module. This module has level 2, and is similar to the highness requirement of Theorem 2.1, modified in the obvious way by replacing K with C. It will be satisfied in exactly the same way as in Theorem 2.1, so needs no further discussion.

5.3.2. The Basic N_e-Module. This module has level 1, with a single initial node α having terminal immediate successor α_0 (α_1, resp.) along which α is activated (validated, resp.) and has Π outcome (Σ outcome, resp.). Clauses (i)–(iii) of Definition 2.3.1 follow easily. As $\text{lev}(\alpha) = 1$, we can interpret α as $\eta^1 \in T^1$ when η^1 is an α–node.

Directing sentence S_α: $\exists u \exists s \in V_\alpha(\Phi_e(A \upharpoonright u; e)[s] \downarrow)$.
Activated action for α at $\rho \supseteq \alpha_0$: $\exists r \forall t \geq r(\Theta(\chi_{\alpha_0}, \gamma_{\text{out}(\alpha_0)}; e)[t] \downarrow = 0)$.
Validated action for α at $\rho \supseteq \alpha_1$:

$$\exists r \forall t \geq r \forall q \geq s(\Theta(\chi_{\alpha_1}, \gamma_{\text{out}(\alpha_1)}; e)[t] \downarrow = 1 \;\&\; A^q \upharpoonright u = A^s \upharpoonright u).$$

The numbers s and u witnessing the truth of the directing sentence S_α are used as parameters for validated action. It is easily checked that this description of directing sentence and action imply the satisfaction of N_e.

5.3.3. T^2-Analysis. The standard initial assignment process is followed, so clause (i) of Definition 2.7.2 follows from Lemma 2.7.4 (Standard Initial Assignment). Hence it suffices to show that all requirements are satisfied.

The standard initial specification for $\eta^2 \in T^2$ is followed, with action supported at $\{\rho^2 \supset \eta^2 : \eta^2 \text{ is } \rho^2\text{-free}\}$. Clause (i) of Definition 2.10.1 will follow from the analysis for the basic modules and the comments preceding the theorem, once we show, below, that χ_{α_1} is a substring of C.

5.3.4. T^1-Analysis. The standard derived assignment is followed. Clauses (i)–(iii) of Definition 2.7.7 follow from Lemma 2.12.2 (Standard Derived Assignment).

Sentence decomposition for β-nodes (Q_e requirements) is as in Section 5.2, and is not required on T^1 for α-nodes. Action for $\eta^1 \in T^1$ is supported at $\{\rho^1 \supset \eta^1 : \eta^1 \text{ is } \rho^1\text{-free}\}$. By Lemma 3.1.8(i,ii) ($\Sigma_1$ Analysis), the construction has infinite support on T^1 and support propagates from T^1 to T^2.

5.3.5. T^0-Analysis. The standard derived assignment is followed. Clauses (i)–(iii) of Definition 2.7.7 follow from Lemma 2.12.2 (Standard Derived Assignment).

Fix $\eta = \nu^- \subset \nu \subseteq \rho \in T^0$. Sentence decomposition for β-nodes is the same as for highness requirements, and is not repeated here. The bounds for the sentence decomposition for α-nodes is as follows, when action for η is supported at ρ: $u = \text{wt}(\lambda(\eta))$, $s = \text{wt}(\eta)$, $t, q \leq \text{wt}(\rho)$, and $r = \text{wt}(\nu)$. Note that we may fix the witnesses s and u for α when we specify validated action. Monotonicity of sentence decomposition follows from (2.8.1) and (2.8.4). Action for η is supported at $\{\rho \supset \eta : \eta \text{ is } \rho\text{-free}\}$. By Lemma 3.1.8(i,ii) ($\Sigma_1$ Analysis), the construction has infinite support on T^1 and support propagates from T^0 to T^1.

5.3.6. The Construction. The construction is executed in accordance with clause (iva) of Definition 2.10.1.

5.3.7. The Verification. Let $\Lambda^0 \in [T^0]$ be the true path for the construction, and for all $i \leq 2$, let $\Lambda^i = \lambda^i(\Lambda^0)$. It is easily seen that the construction admits consistent action. Clauses (i)–(iii) of Definition 3.1.7 are routine to verify, so by Lemma 3.1.9(ii) (Restraint), shared set action is compatible. Furthermore, hypothesis (i) of Lemma 5.1.1 is satisfied for the functional Δ, so by Lemma 5.1.1 (Π_2-Restraint), shared functional action for Δ is compatible. Compatibility of shared functional axioms for Θ follows from Lemma 5.1.5 (Incompatible Oracle).

It follows from the correctness of action for nodes along Λ^2, that if $\eta^2 \subset \Lambda^2$ is an α-node, then χ_{η^2} is a substring of C, so is uniformly computable from C. Furthermore by Remark 2.4.2(i) and as $\eta^2 \subset \Lambda^2$, $\text{out}(\eta^2) \subset \Lambda^1$ so is computable from $\emptyset'$. Let $D = \gamma_{\Lambda^2}$. As $C \geq_T \emptyset'$ and $\Lambda^1 \leq_T \emptyset'$, $\Theta(C, D) = A'$, so clause (i) of Definition 2.10.1 holds. Thus we can apply Lemma 3.1.6 to conclude that clause (iv) of Definition 2.10.1 holds. The theorem follows from Theorem 2.11.5 (Modified Framework). $\dashv$

5.4. The Minimal Pair Theorem

The Minimal Pair Theorem, originally proved independently by Lachlan [7] and Yates [27], asserts the existence of a *minimal pair* of non-computable computably enumerable degrees, i.e., a pair of incomparable computably enumerable degrees whose greatest lower bound is $\boldsymbol{0}$. The proof of the theorem requires the preservation of infinitely many axioms in order to satisfy a single requirement, and so the construction will require level 2 requirements. The compatibility of shared functional axioms will require us to use a *preferential derived assignment* of requirements rather than the standard derived assignment on T^1 and T^0.

Before proving the Minimal Pair Theorem, we prove a lemma that highlights the main idea used to satisfy meet preservation requirements. The lemma is stated just for minimal pair requirements, i.e., requirements that ensure that the meet of a pair of computably enumerable degrees is $\boldsymbol{0}$. A later remark indicates how to extend this lemma to make it applicable for the situation in which the meet of the computably enumerable degrees under construction is not $\boldsymbol{0}$. Roughly, the lemma states the following. Suppose that $\eta_0 \subset \eta_1 \subset \Lambda^0$ are given such that $\Phi_0(A_0 \restriction u_\xi; x)[\mathrm{wt}(\xi)] = \Phi_1(A_1 \restriction u_\xi; x)[\mathrm{wt}(\xi)]$ for $\xi \in \{\eta_0, \eta_1\}$, but that this equality fails for all $\xi \in (\eta_0, \eta_1)$. Then if, for some $i \leq 1$, no number enters A_i due to action at any $\xi \in (\eta_0, \eta_1]$, then the values for the axioms computed at both η_0 and η_1 are identical.

LEMMA 5.4.1 (Meet Preservation Lemma). *Fix $\eta_0 \subset \eta_1 \subset \Lambda^0$ such that the following conditions hold*:

(*i*) *For all $\xi \in \{\eta_0, \eta_1\}$, $\Phi_0(A_0 \restriction u_\xi; x)[\mathrm{wt}(\xi)] = \Phi_1(A_1 \restriction u_\xi; x)[\mathrm{wt}(\xi)]$.*

(*ii*) *$A_i^{\mathrm{wt}(\eta_0)} \restriction u_{\eta_0} = A_i^{\mathrm{wt}(\eta_1)} \restriction u_{\eta_0}$ for some $i \leq 1$.*

Then $\Phi_i(A_i \restriction u_{\eta_0}; x)[\mathrm{wt}(\eta_0)] = \Phi_i(A_i \restriction u_{\eta_1}; x)[\mathrm{wt}(\eta_1)]$ for $i \leq 1$.

PROOF. Fix i as in (ii), so that $A_i^{\mathrm{wt}(\eta_0)} \restriction u_{\eta_0} = A_i^{\mathrm{wt}(\eta_1)} \restriction u_{\eta_0}$. By the consistency of computations and (i),

$$\begin{aligned}\Phi_i(A_i \restriction u_{\eta_0}; x)[\mathrm{wt}(\eta_0)] &= \Phi_i(A_i \restriction u_{\eta_0}; x)[\mathrm{wt}(\eta_1)] \\ &= \Phi_{1-i}(A_{1-i} \restriction u_{\eta_1}; x)[\mathrm{wt}(\eta_1)].\end{aligned}$$ ⊣

Lemma 5.4.1 is used in the following way. Clause (ii) of Lemma 5.4.1 will be imposed on every pair of nodes $\eta_0 \subset \eta_1$ such that $\mathrm{up}^2(\eta_0) = \mathrm{up}^2(\eta_1)$, there is no node ξ between η_0 and η_1 such that $\mathrm{up}^2(\xi) = \mathrm{up}^2(\eta_0)$, and a meet requirement is assigned to η_0. Thus the first common value m for $\Phi_0(A_0 \restriction u_\xi; x)[\mathrm{wt}(\xi)]$ and $\Phi_1(A_1 \restriction u_\xi; x)[\mathrm{wt}(\xi)]$ computed for some ξ such that $\mathrm{up}^2(\xi) = \mathrm{up}^2(\eta_0)$ is the value of $\Phi_0(A_0; x)$ whenever $\Phi_0(A_0; x) = \Phi_1(A_1; x)$. This provides an effective computation of $\Phi_0(A_0; x)$.

REMARK 5.4.2. The condition that $i \leq 1$ can be replaced, in the obvious way by $i \leq m$ for any $m \in \mathbb{N}$. Also, for non-zero meets, if the functional $\Delta(B; x)$ is computing the common values in (i) and has use v_{η_0} at η_0, then one can replace (ii) by a condition that the failure of (ii) as stated will imply that $B^{\mathrm{wt}(\eta_0)} \restriction v_{\eta_0} \neq B^{\mathrm{wt}(\eta_1)} \restriction v_{\eta_0}$.

We now describe a situation that precludes our use of the standard derived assignment on T^1 and T^0; the problems encountered will be resolved by the preferential derived assignment, defined below.

Suppose that $\eta^2 \subset \xi^2 \subset \sigma^2 \in T^2$, η^2 is assigned a minimal pair requirement trying to preserve computations from oracles A_0 and A_1, ξ^2 is assigned a non-computability requirement whose action places a number into A_0, and σ^2 is assigned a non-computability requirement whose action places a number into A_1. Suppose further that $\nu \subset \Lambda^0$, $\nu^- = \sigma$, that $\xi^1 \subset \sigma^1 \subset \eta^1 \subset \lambda(\sigma)$ with $\text{up}(\xi^1) = \xi^2$, $\text{up}(\sigma^1) = \sigma^2$ and $\text{up}(\eta^1) = \eta^2$, and that ν switches σ^1. Let μ^1 be the initial derivative of η^2 along η^1, and note that by (2.6.1) and Definition 2.4.1, $\mu^1 \subset \xi^1$. Then an axiom $\Delta(\emptyset; x) = m$ may have been declared for $x = \text{wt}(\sigma^1)$, action for σ at ν may injure the corresponding axiom $\Psi(A_1; x) = m$ by placing a number into A_1, but no node $\subseteq \sigma^1$ has directing sentence that checks if $\Phi(A_0; x) = \Psi(A_1; x)$. Hence there is no way to prevent almost immediate action for ξ^1 that would place a number into A_0 and thereby injure the axiom $\Phi(A_0; x) = m$ before we have a recovery for the axiom $\Psi(A_1; x) = m$, if the standard derived assignment is followed. The solution will be to use the preferential derived assignment procedure, which we now define, to prevent such a sequence of events.

DEFINITION 5.4.3. The *preferential derived assignment* at level k is defined as follows. Suppose that we are given a partition of the set $\mathcal{R}$ of all requirements into two well-ordered sets of order-type ω, $\mathcal{R}_0$ and $\mathcal{R}_1$, with $\mathcal{R}_0$ specified as the *preferred class*. We divide a block along T^k into *subblocks*, begin a subblock either at the node at which the block begins or at a node to which a requirement in $\mathcal{R}_1$ is assigned, and end a subblock along ρ^k at ν^k if there is a $\tau^k \subseteq \rho^k$ such that $(\tau^k)^- = \nu^k$ and either ν^k is the initial derivative of a node ending a block along $\lambda(\tau^k)$, or a requirement in $\mathcal{R}_1$ is assigned to τ^k. We specify that the antiderivative η^{k+1} of η^k is the shortest η^k-consistent node such that either:

(i) a requirement in $\mathcal{R}_0$ is assigned to η^{k+1}, and η^{k+1} has an antiderivative $\subset \eta^k$, but none in the subblock containing η^k; or
(ii) there is no node satisfying (i) and η^{k+1} has no derivative in the block containing η^k.

The preferential derived assignment circumvents the problem mentioned in the paragraph preceding Definition 5.4.3; it is used both on T^1 and on T^0. All subblocks on T^1 will be full blocks, and a new block will begin immediately following each node $\eta^1 \in T^1$ to which a requirement $P_{\Phi,i}$ is assigned; henceforth, we refer to such nodes as α-nodes. Every path through this block will begin with derivatives of those nodes $\sigma^2 \subset \text{up}(\eta^1)$ that have Π outcome along $\text{up}(\eta^1)$ and to which a requirement $N_{\tilde{\Phi},\Psi}$ has been assigned (henceforth we call such nodes β-nodes) until either we reach a β-node of T^1 with Π outcome, or until all such nodes have derivatives in the block. Each derivative $\sigma^1 \in T^1$ of such a σ^2 will check for a longer length of agreement than previously encountered during the construction, and all successors of such σ^1 will have weight larger than the length of agreement being checked for, so cannot injure a computation witnessed at σ^1. On T^0, a new subblock will begin immediately following each α-node $\eta^0 \in T^0$. This subblock will begin with derivatives of the β-nodes σ^0 through the current path through T^1, so in particular, will place derivatives σ^0 of the β-nodes σ^1 just mentioned before any derivative of an α-node. If such a σ^0 is activated, then its antiderivative $\sigma^1 \in T^1$ will have Π outcome and will not be an initial derivative, so will form the end of a link around any α-node of T^1 whose antiderivative on T^2 lies in $[\mathrm{up}(\sigma^1), \lambda(\sigma^1)]$, thereby preserving one correct computation for σ^1 until a sufficiently long length of agreement is again observed.

REMARK 5.4.4. It is easy to see that any preferential derived assignment satisfies clause (i) of Definition 2.7.7. We will show in Lemma 8.6.2 that this assignment also satisfies clause (ii) and (iii) of Definition 2.7.7.

The following Lemma will be proved as Lemma 8.6.6, and will be used to show that the hypotheses of Lemma 5.4.1 (Meet Preservation) hold.

LEMMA 5.4.5. *Suppose that the preferential derived assignment is followed on* T^0, *and let* $\mathcal{R}_0$ *be the preferred set of requirements and let* $\mathcal{R}_1$ *be its complement. Let* $\eta_i \subset \nu_i \in T^0$ *for* $i \leq 1$ *be given such that* η_i *is validated along* ν_i, $(\nu_i)^- = \eta_i$, $\nu_0 \subset \nu_1$, $\mathrm{up}^2(\eta_0) = \mathrm{up}^2(\eta_1)$, *and there is no* $\xi \in (\eta_0, \eta_1)$ *such that* $\mathrm{up}^2(\xi) = \mathrm{up}^2(\eta_0)$ *and* ξ *is validated along* η_1. *Then there is at most one* $\sigma \in [\nu_0, \nu_1)$ *such that a requirement in* $\mathcal{R}_1$ *is assigned to* σ, $\mathrm{wt}(\mathrm{up}(\sigma)) \leq \mathrm{wt}(\mathrm{up}(\eta_0))$, *and* σ *is validated along* ν_1.

THEOREM 5.4.6 (Minimal Pair Theorem). *There are computably enumerable degrees* $\boldsymbol{a_0}, \boldsymbol{a_1} > \boldsymbol{0}$ *such that* $\boldsymbol{a_0} \cap \boldsymbol{a_1} = \boldsymbol{0}$.

PROOF. We construct computably enumerable sets A_0 and A_1 whose degrees form a minimal pair. The following level 1 requirement 1 will be satisfied for each computable partial functional Φ and each $i \leq 1$:

$$P_{\Phi,i} : \Phi(\emptyset) \neq A_i.$$

And for each pair of computable partial functionals $\langle \Phi, \Psi \rangle$, we construct a computable partial functional Δ to satisfy the following level 2 requirement:

$$N_{\Phi,\Psi} : \Phi(A_0) = \Psi(A_1) \;\&\; \Phi(A_0) \text{ total } \rightarrow \exists\Delta(\Delta(\emptyset) = \Phi(A_0)).$$

Let $\{R_i : i \in \mathbb{N}\}$ be an effective list of all requirements.

5.4.1. The Basic $P_{\Phi,i}$-Module. These requirements are the non-computability requirements introduced in Section 3.3. Although the construction begins on T^2, (2.6.3) pulls these requirements down to T^1, where parameters are interpreted. The treatment of these requirements is thus identical to that in Section 3.3, so is not repeated in this section.

5.4.2. The Basic $N_{\Phi,\Psi}$-Module. This module consists of a single non-terminal node β having level 2, with two terminal extensions β_0 and β_1; β is activated (validated, resp.) and has Σ outcome (Π outcome, resp.) along β_0 (β_1, resp.). Clauses (i)–(iii) of Definition 2.3.1 follow easily. These is no activated action for β. The validated action sentence for β will use parameters $\beta^0_{\beta^1}$ and u_{β^1}; these parameters will be witnesses to the validity of S_{β^1}.

Directing sentence S_β:

$$\forall \beta^1 \in V_\beta \exists \beta^0 \in V_{\beta^1} \forall y \leq \mathrm{wt}(\beta^1) \exists u$$
$$(\Phi(A_0 \restriction u; y)[\mathrm{wt}(\beta^0)] \downarrow = \Psi(A_1 \restriction u; y)[\mathrm{wt}(\beta^0)] \downarrow).$$

Validated action for β at $\rho \supseteq \beta_1$:

$$\forall \beta^1 \in V_\beta \exists r \forall y \leq \mathrm{wt}(\beta^1) \exists v \forall t \geq r$$
$$(\Delta_\beta(\emptyset; y)[r] \downarrow = \Phi(A_0 \restriction u_{\beta^1}; y)[\beta^0_{\beta^1}] \;\&$$
$$(\Phi(A_0 \restriction v; y)[t] = \Phi(A_0 \restriction u_{\beta^1}; y)[\beta^0_{\beta^1}] \text{ or}$$
$$\Psi(A_1 \restriction v; y)[t] = \Psi(A_1 \restriction u_{\beta^1}; y)[\beta^0_{\beta^1}])).$$

If S_β is true and validated action is followed at extensions of β, then $\Phi(A_0) = \Psi(A_1) = \Delta(\emptyset)$; and if S_β is false, then $\Phi(A_0) \neq \Psi(A_1)$. In either case, we see that the above action will ensure that $N_{\Phi,\Psi}$ is satisfied.

5.4.3. *T^2*-Analysis. The standard initial assignment is followed, so clause (i) of Definition 2.7.2 follows from Lemma 2.7.4 (Standard Initial Assignment). Thus it suffices to show that all requirements are satisfied.

The standard initial sentence specification is followed, and action for $\eta^2 \in T^2$ is supported at $\{\rho^2 \supset \eta^2 : \eta^2 \text{ is } \rho^2\text{-free}\}$. Clause (i) of Definition 2.10.1 follows from the analysis of basic modules.

5.4.4. *T^1*-Analysis. The preferential derived assignment is followed, with preference given to the class of requirements of the form $N_{\Phi,\Psi}$. Clauses (i)–(iii) of Definition 2.7.7 follow from Lemma 8.6.2 (Preferential Derived Assignment).

Sentence decomposition for β-nodes at $\eta^1 \in T^1$ specifies $\beta^1 \subseteq \eta^1$. Monotonicity of sentence decomposition follows from (2.8.1). Action for $\eta^1 \in T^1$ is supported at $\{\rho^1 \supset \eta^1 : \eta^1 \text{ is } \rho^1\text{-free}\}$. By Lemma 3.1.8(i,ii) ($\Sigma_1$ Analysis), the construction has infinite support on T^1 and support propagates from T^1 to T^2.

5.4.5. *T^0*-Analysis. The preferential derived assignment is followed, with preference given to the class of requirements of the form $N_{\Phi,\Psi}$. Clauses (i)–(iii) of Definition 2.7.7 follow from Lemma 8.6.2 (Preferential Derived Assignment).

Fix $\eta = \nu^- \subset \nu \subseteq \rho \in T^0$. Sentence decomposition for β-nodes is as follows. The bounds for the sentences are: $u = \mathrm{wt}(\lambda(\eta))$, $\beta^0 \subseteq \eta$, $v = \mathrm{wt}(\lambda(\rho))$, $r = \mathrm{wt}(\nu)$, and $t \leq \mathrm{wt}(\rho)$. Monotonicity of sentence decomposition follows from (2.8.1) and (2.8.4). Action for η is supported at $\{\rho \supset \eta : \eta \text{ is } \rho\text{-free}\}$. By Lemma 3.1.8(i,ii) ($\Sigma_1$ Analysis), the construction has infinite support on T^0 and support propagates from T^0 to T^1.

5.4.6. The Construction. The construction is executed in accordance with clause (iva) of Definition 2.10.1.

5.4.7. The Verification. Let $\Lambda^0 \in [T^0]$ be the true path for the construction, and for $i \leq 2$, let $\Lambda^i = \lambda^i(\Lambda^0)$. It is easily checked that the construction admits consistent action. We note that clauses (i)–(iii) of Definition 3.1.7 hold, so by Lemma 3.1.9(ii) (Restraint), shared set axioms are compatible. Suppose that $\langle \eta_0, \rho_0 \rangle$ and $\langle \eta_1, \rho_1 \rangle$ share a functional axiom with argument x. It follows from the sentence decomposition that $x \leq \text{wt}(\text{up}(\eta_0)), \text{wt}(\text{up}(\eta_1))$. Then $\text{up}^2(\eta_0) = \text{up}^2(\eta_1)$ and for $i \leq 1$, η_i must be validated along ρ_i as functional axioms are declared only in this situation. We may assume that $(\rho_i)^- = \eta_i$ for $i \leq 1$, as new instructions to declare functional axioms only occur in this situation. We may also assume without loss of generality that there is no $\sigma \in (\eta_0, \eta_1)$ such that $\text{up}^2(\sigma) = \text{up}^2(\eta_0)$ and σ is validated along ρ_1, else we could replace η_1 with σ and proceed by induction. Hence we can apply Lemma 5.4.5 and clause (ii) of Definition 3.1.7 to conclude that the hypotheses of Lemma 5.4.1 (Meet Preservation) hold. It now follows from Lemma 5.4.1 (Meet Preservation) that these shared functional axioms are compatible.

We now note that Lemma 3.1.6 can be applied to conclude that clause (iv) of Definition 2.10.1 holds. The theorem now follows from Theorem 2.11.5 (Modified Framework). $\dashv$

5.4.8. Further Remarks. It is not difficult to see that the highness requirements of Section 5.2 can be combined with the alternating restraint algorithm of this section. The preferential derived assignment is implemented on T^1 and T^0, and the preferred class of requirements is the class of minimal pair preservation requirements. We leave the details of the proof of the following theorem to the reader.

THEOREM 5.4.7. *There is a minimal pair of high degrees.*

The embedding of countable distributive lattices into the computably enumerable degrees due to Thomason [26] and independently to Lerman (cf. [25]) (that provides a construction of a computably enumerable nonzero branching degree B, i.e., a computably enumerable degree that is the g.l.b. of two larger computably enumerable degrees), has a similar proof. The only difference in the proof is in the satisfaction of a requirement $\Phi(B) \neq A_i$; there may be requirements that place a number x into the set B, thereby injuring computations from both A_i and A_{1-i} for an infimum requirement $N_{\Phi,\Psi}$. But then $B \leq_T A_i, A_{1-i}$ and it will be the case that x is less than the use of the axiom $\Delta_\tau(B; x) = m$ in existence at the time, so this axiom will be destroyed, and a new axiom can be declared when needed.

5.5. Embedding the Pentagon

The techniques introduced to prove Theorem 5.4.6 (Minimal Pair) can readily be extended to embed all countable distributive lattices into the computably enumerable degrees preserving least element. When trying to embed nondistributive lattices, however, the conflicts between requirements become more severe. We look at the simplest such case next, namely, we embed the lattice N_5 into the computably enumerable degrees (a result of Lachlan [La3]). This lattice has five elements: a smallest element 0, a largest element d, comparable elements $c < b$ in $(0, d)$ and an element $a \in (0, d)$ that is incomparable with both b and c. The proof uses an expanded basic module that ensures that the set B representing b is not computable from the set C representing c.

THEOREM 5.5.1. *There are computably enumerable sets A, B and C whose degrees generate a lattice isomorphic to N_5.*

PROOF. It is readily seen that it suffices to construct computably enumerable sets A, B, and C that satisfy the following requirements:

$$L : C = \Gamma(B).$$
$$J : B = \Delta(A, C).$$
$$P_{A,\Phi} : A \neq \Phi(\emptyset).$$
$$P_{C,\Phi} : C \neq \Phi(\emptyset).$$
$$Q_\Phi : B \neq \Phi(C).$$
$$N_{\Psi,\Theta} : \Psi(A) = \Theta(B) \ \&\ \Psi(A) \text{ total} \to \exists \Xi(\Psi(A) = \Xi(\emptyset)).$$

We order all requirements except L and J in an effective list $\{R_i : i \in \mathbb{N}\}$. L and J will be seen to be satisfied through general properties of the construction, and need not be placed on a tree of strategies.

5.5.1. The L Requirement. We ensure that $\Gamma(B; x) = C(x)$ for all x, by forcing the construction to satisfy the following property for every s:

$$x \in C^{s+1} - C^s \to x \in B^{s+1} - B^s. \tag{5.5.1}$$

Since the use of any axiom on argument x is x, it is easily seen that (5.5.1) implies the satisfaction of the L requirement.

5.5.2. The J Requirement. We ensure that $\Delta(A, C; x) = B(x)$ for all x, in much the same way as for the L requirement, except that we cannot bound the use for computations in advance, so cannot always place the same number into A or C as was placed into B in order to correct computations. A number x will require a two-step process to enter B if it is not simultaneously placed into C. The first step, taking place at a stage r, will be a decision to place $x \in A^r - A^{r-1}$, and to appoint a trace $t(x)$ targeted for C. The second step, taking place at stage $s > r$, will be to place $x \in B^s - B^{s-1}$ and simultaneously to place $t(x) \in C^s - C^{s-1}$. As the decision to take each step is done computably, this will provide a computable computation of B from $A \oplus C$. Thus it will suffice to show that the construction has the following property for every s:

$$x \in B^s - B^{s-1} \leftrightarrow (x \in C^s - C^{s-1} \vee \exists r < s(x \in A^r - A^{r-1} \ \&\ t(x)[r]\downarrow \ \&\ t(x)[r] \in C^s - C^{s-1})). \quad (5.5.2)$$

5.5.3. The Basic $N_{\Psi,\Theta}$-Module. This is the minimal pair module of Theorem 5.4.6, with the same directing sentence and action.

5.5.4. The Basic $P_{A,\Phi}$-Module and the Basic $P_{C,\Phi}$-Module. These are similar to the non-computability modules of Section 3.3, with the following added action for $P_{C,\Phi}$; whenever we place $\mathrm{wt}(\eta^1)$ into C, we simultaneously place $\mathrm{wt}(\eta^1)$ into B.

5.5.5. The Basic Q_Φ-Module. A first attempt would be to try to use the incomparability requirements of Theorem 3.2.1 (Friedberg–Mučnik). This can be done, if we hold the J requirement responsible for its own satisfaction. It is simpler, however, to transfer this responsibility to the Q_Φ requirement. The way that this is done was discussed earlier, but we repeat that discussion here.

In order to place an element x into B, we must also place a number into A or C in order to keep the J-requirement correct. We cannot place such a number into C, as this action would injure the directing sentence for Q_Φ. We also cannot place numbers into A and B simultaneously, as such action would injure previous action taken for $N_{\Psi,\Theta}$-requirements that we will be unable to correct. Instead, we follow a two-step process. We first place a number into A only, allowing us to revise the axiom for $\Delta(A, C; x)$, giving it a new use $y = t(x)$ that is larger than the use of $\Phi(C; x)$. This action will force the true path to change, as some $N_{\Psi,\Theta}$-requirements will be injured. If the node returns to the true path, then we will be able to simultaneously place x into B and y into C, allowing us to satisfy both Q_Φ and J.

This basic module has two non-terminal nodes γ and γ_1, each of level 1, and three terminal nodes γ_0, γ_{10}, and γ_{11}. Each of the non-terminal nodes completes a block along all extensions. Outcomes for γ and γ_1 along their successors are determined in the standard way. The satisfaction of clauses (i)–(iii) of definition 2.3.1 follow easily.

Care must be taken when the module is placed on T^1, as we may interpolate other nodes between the interpretation of γ and that of γ_1. Thus we differentiate between the two versions of γ_1, using γ_1 to represent the node at which the second part of the module begins to act, and $\widetilde{\gamma}_1$ to represent the immediate successor of γ along which γ is validated. Parameters for the sentences for γ are interpreted at nodes $\gamma^1 \in T^1$ such that $\mathrm{up}(\gamma^1) = \gamma$, with $\widetilde{\gamma}_1^1$ interpreting $\widetilde{\gamma}_1$; and parameters for the sentences for γ_1 are interpreted at nodes $\gamma_1^1 \in T^1$ such that $\mathrm{up}(\gamma_1^1) = \gamma_1$. Note that the validated action sentence for γ_1 uses $\mathrm{wt}(\widetilde{\gamma}_1^1)$, not $\mathrm{wt}(\gamma_1^1)$ as its trace for C. There is no activated action for γ_1.

Directing sentence S_γ: $\exists u \exists s \in V_{\gamma^1}(\Phi(C \restriction u; \mathrm{wt}(\gamma^1))[s] \downarrow = 0)$.
Activated action for γ at $\rho \supseteq \gamma_0$: $\forall t(\mathrm{wt}(\gamma^1) \notin B^t \cup A^t \cup C^t)$.
Validated action for γ at $\rho \supseteq \widetilde{\gamma}_1$:

$$\exists r \forall t \geq s(\mathrm{wt}(\gamma^1) \in A^r \ \& \ C^t \restriction u = C^s \restriction u).$$

Directing sentence S_{γ_1}: $0 = 0$.
Validated action for γ_1 at $\rho \supseteq \gamma_{11}$: $\exists r(\mathrm{wt}(\gamma^1) \in B^r \ \& \ \mathrm{wt}(\widetilde{\gamma}_1^1) \in C^r)$.

If S_γ is false and activated action is followed for γ, then there is an x such that either $\Phi(C; x) \uparrow$, or $\Phi(C; x) \downarrow \neq 0 = B(x)$; and if S_γ is true and validated action for γ is followed at $\rho \supseteq \gamma_{11}$, then there is an x such that $\Phi(C; x) \downarrow = 0 \neq 1 = B(x)$. In either case, the requirement is satisfied. The other clauses of the action sentence are needed to show that (5.5.2) is satisfied.

5.5.6. T^2-Analysis. The standard initial assignment is followed. Clause (i) of Definition 2.7.2 follows from Lemma 2.7.4 (Standard Initial Assignment). Thus it suffices to show that all requirements are satisfied.

The standard initial sentence specification is followed, and action for $\eta^2 \in T^2$ is supported at $\{\rho^2 \supset \eta^2 : \eta^2 \text{ is } \rho^2\text{-free}\}$. Clause (i) of Definition 2.10.1 follows from the analysis of basic modules.

5.5.7. T^1-Analysis. The preferential derived assignment is followed, with preference given to the class of requirements of the form $N_{\Psi,\Theta}$. Clauses (i)–(iii) of Definition 2.7.7 follow from Lemma 8.6.2 (Preferential Derived Assignment).

Sentence decomposition for the requirements Q_Φ will not be required until we reach T^0, except that the parameters γ and γ_1 are interpreted as the nodes of T^1 to which the given nodes of the basic module are assigned. Sentence decomposition for the other requirements has been described in previous sections. Monotonicity of sentence decomposition follows from (2.8.1). Action for $\eta^1 \in T^1$ is supported at $\{\rho^1 \supset \eta^1 : \eta^1 \text{ is } \rho^1\text{-free}\}$. By Lemma 3.1.8(i,ii) ($\Sigma_1$ Analysis), the construction has infinite support on T^1, and support propagates from T^1 to T^2.

5.5.8. T^0-Analysis. The preferential derived assignment is followed, with preference given to the class of requirements of the form $N_{\Psi,\Theta}$. Clauses (i)–(iii) of Definition 2.7.7 follow from Lemma 8.6.2 (Preferential Derived Assignment).

Sentence decomposition for requirements $P_{A,\Phi}$ and $P_{C,\Phi}$ is the standard decomposition for non-computability requirements introduced in Section 3.3, and sentence decomposition for $N_{\Psi,\Theta}$ is the same as that for the minimal pair requirements of Section 5.4. Fix $\eta = \nu^- \subset \nu \subseteq \rho \in T^0$. Sentence decomposition for γ- and γ_1-nodes at η is as follows. $u \leq \mathrm{wt}(\lambda(\eta))$, $s = \mathrm{wt}(\eta)$, $r = \mathrm{wt}(\nu)$ and $t \leq \mathrm{wt}(\rho)$. Monotonicity of the sentence decomposition follows from

(2.8.1) and (2.8.4). Action for η is supported at $\{\rho \supset \eta : \eta \text{ is } \rho\text{-free}\}$. By Lemma 3.1.8(i,ii) ($\Sigma_1$ Analysis), the construction has infinite support on T^0, and support propagates from T^0 to T^1.

5.5.9. The Construction. The construction is executed in accordance with clause (iva) of Definition 2.10.1.

5.5.10. The Verification. Let $\Lambda^0 \in [T^0]$ be the true path for the construction, and for $i \leq 2$, let $\Lambda^i = \lambda^i(\Lambda^0)$. It is easily checked that the construction admits consistent action. Note that the preferential derived assignment on T^1 will interpolate only level 2 nodes between a γ-node and a γ_1-node derived from the same fixed interpretation of a basic module on T^2. Hence clauses (i) and (ii) of Definition 3.4.1 hold, so by Lemma 3.4.2(ii) (Delayed Set Action), shared set axioms are compatible.

Suppose that $\langle \eta_0, \rho_0 \rangle$ and $\langle \eta_1, \rho_1 \rangle$ share a functional axiom with argument x. It follows from the sentence decomposition that $x \leq \text{wt}(\text{up}(\eta_0)), \text{wt}(\text{up}(\eta_1))$. Then $\text{up}^2(\eta_0) = \text{up}^2(\eta_1)$ and for $i \leq 1$, η_i must be validated along ρ_i as functional axioms are declared only in this situation. We may assume that $(\rho_i)^- = \eta_i$ for $i \leq 1$, as new instructions to declare functional axioms only occur in this situation. We may also assume without loss of generality that there is no $\sigma \in (\eta_0, \eta_1)$ such that $\text{up}^2(\sigma) = \text{up}^2(\eta_0)$ and σ is validated along ρ_1, else we could replace η_1 with σ and proceed by induction. Hence we can apply Remark 5.4.5 and clause (ii) of Definition 3.1.7, noting that action for a γ_1-node will leave either the oracle for Φ uninjured or the oracle for Θ uninjured for the requirement $N_{\Phi,\Theta}$, to conclude that the hypotheses of Lemma 5.4.1 (Meet Preservation) hold. It now follows from Lemma 5.4.1 (Meet Preservation) that these shared functional axioms are compatible.

Our final observation is that for requirements Q_Φ, the parameter $\text{wt}(\tilde{\gamma}_1^1)$ is appointed immediately upon the entry of $\text{wt}(\gamma_1)$ into A, so (5.5.2) holds for a computable parameter t. Thus the J requirements are satisfied. It is easy to see that the L requirements are satisfied.

We now note that Lemma 3.1.6 can be applied to conclude that clause (iv) of Definition 2.10.1 holds. The theorem now follows from Theorem 2.11.5 (Modified Framework). ⊣

5.5.11. Further Remarks. The *trace-probe property* introduced by Lerman, Shore and Soare [17] is a property that generalizes the techniques used to embed the pentagon. One requires the basic module to be finite, with a number entering a set at each step, thereby allowing the change of target set for some join requirement. At the end, one reaches a stable situation in which all numbers can enter their target sets simultaneously to satisfy the diagonalization requirement. Such a construction can be carried out using the framework introduced in this book.

We note that the construction presented in this section differs from the previous constructions in that we use external conditions to show that requirements are satisfied. In particular, L and J requirements delegate the responsibility for their satisfaction to other requirements. This is not necessary, but greatly simplifies the proof. Such a delegation of responsibility will not succeed for more complicated lattice embedding constructions. What is required in those cases is a much more delicate and careful assignment of requirements to T^1 and T^0.

5.6. Summary

We saw several new ways to ensure that shared functional action is compatible in this chapter. In Sections 5.2 and 5.3, we were able to ensure that axioms declared by incomparable nodes on T^2 used incomparable oracles. Axioms declared for the same nodes of T^2 and of T^1 were corrected when the action changed the value for the axiom, and were mutually compatible if the nodes of T^1 were different and both nodes on T^1 were free. In Sections 5.4 and 5.5, we used alternating restraint to prevent the values of axioms from changing.

A more interesting conflict arose in Section 5.5. We were faced with action for a single module violating the alternating restraint necessary in order to show that shared axioms declared for functionals controlled by other nodes were compatible. This forced us to take basic modules with consecutive nodes on T^2, and separate the derivatives of these nodes on T^1 by interpolating nodes imposing the alternating restraint. This is the first time that we were forced to intermix modules, but we will see such a situation occur again at higher levels.

Chapter 6

Δ_3 CONSTRUCTIONS

Δ_3 constructions are similar to Δ_2 constructions, raised by a level. They take place on four trees, T^0, T^1, T^2 and T^3, and are characterized by the following property: If $\eta^3 \subset \nu^3 \in T^3$, ν^3 corresponds to a terminal node of the basic module associated with η^3, $\text{lev}(\eta^3) = 3$, and η^3 has Π outcome along ν^3, then no requirement is assigned to ν^3 or any of its successors. In this case, the falsity of the directing sentence assigned to η^3 together with activated action for this node suffices to prove the theorem.

We prove one theorem falling under this classification, the Sacks Density Theorem, in Section 6.2. We discuss certain types of requirements encountered in Δ_3 constructions and state lemmas delineating properties of these requirements in Section 6.1.

6.1. Δ_3 Constructions

When functional axioms need to be declared and Player I controls the enumeration of one of the oracle sets, the construction becomes more complex. This is the case when we are trying to construct a degree within a given interval. The simplest such case dealing with an arbitrary interval whose endpoints are computably enumerable degrees is the density theorem, proved in the next section; this theorem asserts that there is a computably enumerable set whose degree lies strictly between the endpoints of the interval. Thus we will be given computably enumerable sets $U <_T V$ controlled by Player I, and Player II will be assigned the responsibility to construct a computably enumerable set A such that $U <_T A <_T V$. Player II can replace A by $U \oplus A$, and will use backtracking combined with permitting to show that $A \leq_T V$. Thus the requirements that will need to be satisfied are those stating that the inequalities are proper.

The strategy for satisfying requirements is much like we saw in Chapter 4 where we proved a similar theorem in the case that the smallest element of the interval was the degree of the empty set; however, because we are replacing the empty set with a non-computable set U, we need a construction that is one level higher, namely, a Δ_3^0 construction.

The strategy used to ensure that $V \not\leq_T U \oplus A$ is the restraint strategy used in the proof of Theorem 4.5.2, modified to take changes by the set U into account. A typical requirement of this kind states that if $\Phi(U \oplus A) = V$, then there is a computable partial functional Δ such that $\Delta(U) = V$, thereby implying that it is not the case that $U <_T V$. As we see longer lengths of agreement between $\Phi(U \oplus A)$ and V, Player II tries to restrain A, and simultaneously tries to compute V on the same set of arguments from U alone. If V changes without a corresponding U change, then the requirement is won by Player II as there is a difference between $\Phi(U \oplus A)$ and V. Thus the only way that V can change without presenting Player II with a win is in concert with a U change; but this enables Player II to ensure that $\Delta(U)$ is a correct computation of V. In the case in which $U = \emptyset$, no U changes were possible, so the level of the construction was lower.

The strategy to ensure that $A \not\leq_T U$ is a new strategy. A typical requirement of this kind states that if $\Phi(U) = A$, then there is a computable partial functional Δ such that $\Delta(U \oplus A) = V$, from which it follows that V is computable from a set of degree U and so that the interval is degenerate. If the computation by Δ of V becomes incorrect due to a change to V without a corresponding change to U, then Player II places a number less than or equal to the use of the computation into A, thereby allowing the computation to be corrected. If Player I always reacts to the placement of numbers into A by placing a number into U in order to correct Φ, then it will be the case that $A \leq_T U$, whence the oracle for the computation of V by Δ has degree U, so we will have concluded that $V \leq_T U$.

The implementation of the strategies just described requires the starting tree to be T^3; however, if we reach a node $\eta^3 \in T^3$ at which Δ is total on its oracle and correctly computes V, then the Π outcome of η^3 along the true path is enough to prove the theorem, so no requirements are assigned to proper extensions of η^3 along any of its Π outcomes. Thus the level of the construction is reduced to Δ_3^0. Furthermore, in order for the validity of directing sentences to correspond with the types of outcomes required for a level 3 construction, the directing sentence must be the negation of the sentences mentioned above, so must ask if the length of agreement is *not* sufficiently long.

6.1.1. Properties of Δ_3 Constructions. The proof that the requirements that we encounter are satisfied relies on certain properties of tree decompositions beginning at level 3 and corresponding to Δ_3 Constructions. These lemmas are proved in Section 8.7, but are stated here. The first such lemma deals with the uniqueness of initial derivatives, and is proved as Lemma 8.7.3.

LEMMA 6.1.1. *Let a Δ_3 construction be given with true path $\Lambda^0 \in T^0$. Suppose that $\eta^3 = \lambda^3(\sigma)$ for some $\sigma \subset \Lambda^0$. Then:*

(*i*) *There is a unique* $\xi^2 \in \{\lambda^2(\xi) : \xi \subset \Lambda^0\}$ *such that* ξ^2 *is an initial derivative of* η^3.

(*ii*) *There is a unique* $\xi^1 \in \{\lambda(\xi) : \xi \subset \Lambda^0\}$ *such that* ξ^1 *is an initial derivative of* η^3. ⊣

The next lemma characterizes intervals on T^2 whose endpoints are derivatives of the same node of T^3; it is proved as Lemma 8.7.4.

LEMMA 6.1.2. *Let a* Δ_3 *construction be given. Fix* $\eta^3 \in T^3$ *and* $\xi^2, \eta^2, \kappa^2 \in T^2$ *such that* $\text{up}(\xi^2) = \text{up}(\eta^2) = \eta^3$ *and* $\kappa^2 \subseteq \xi^2 \wedge \eta^2$. *Then*:

(*i*) *If* $\text{up}(\kappa^2) = \eta^3$, *then* $\text{up}(\sigma^2) = \eta^3$ *for all* $\sigma^2 \in [\kappa^2, \eta^2]$.

(*ii*) *If* $\kappa^2 = \xi^2 \wedge \eta^2$, *then* $\text{up}(\kappa^2) = \eta^3$. ⊣

Intervals on T^1 whose endpoints are derivatives of the same node of T^3 are now characterized. This result is proved as Lemma 8.7.5

LEMMA 6.1.3. *Let a* Δ_3 *construction be given. Fix* $\eta^3 \in T^3$ *with initial derivative* $\eta^1 \subset \tau^1 \in \{\lambda(\sigma) : \sigma \subset \Lambda^0\}$ *such that* $\text{up}^3(\eta^1) = \text{up}^3(\tau^1) = \eta^3$. *Then there is a number k and* $\mu_0^1 \subseteq \pi_0^1 \subset \mu_1^1 \subseteq \pi_1^1 \subset \cdots \subset \mu_k^1 \subseteq \pi_k^1$ *such that all* $\xi^1 \in (\pi_i^1, \mu_{i+1}^1)$ *have* Σ *outcome along* μ_{i+1}^1 *for all* $i < k$, $\text{up}(\mu_i^1) = \text{up}(\pi_i^1)$ *for all* $i \leq k$, $(\text{up}(\mu_{i+1}^1))^- = \text{up}(\mu_i^1)$ *for all* $i < k$, π_i^1 *has* Π *outcome along* μ_{i+1}^1 *for all* $i < k$, $\mu_0^1 = \eta^1$ *and* $\pi_k^1 = \tau^1$. ⊣

6.1.2. The Restraint Requirements. The strategy to ensure that $V \not\leq_T U \oplus A$ will consist of showing that a failure to satisfy this condition will provide the ability to define a computable partial functional Θ such that $\Theta(U) = V$. Thus for each computable partial functional Φ, we will satisfy the requirement specifying that if $\Phi(A, U) = V$, then there is a Θ such that $\Theta(U) = V$. Each node of T^3 to which such a requirement is assigned will define its own functional Θ. The strategy will be to restrain A whenever we find a new length of agreement x between $\Phi(A, U)$ and V, and declare axioms $\Theta(U; y) = V(y)$ for $y \leq x$ with U-use at least as large as that of Φ. If $U <_T V$, then the permanent restraint on A will be finite, allowing other requirements to act.

We now describe the implementation of the strategy to satisfy this requirement when it is assigned to $\eta^3 \in T^3$. As the nodes of T^2 working on this requirement increase in length, we look for longer and longer lengths of agreement between $\Phi(A, U)$ and V; the uses obtained for these lengths of agreement will thus be non-decreasing as the length of the node increases. The declaration of axioms for $\Theta(U)$ at a node η^1 with use u will obey clause (i) of Definition 3.1.7 and axioms will be declared only as activated action, and requirements placing numbers into A will obey clause (ii) of Definition 3.1.7. Hence by Lemma 3.1.9(i) (Restraint), shared functional action for Θ will be compatible unless some $\xi^1 \subset \eta^1$ acts to place a number into A after the axiom is declared as action for η^1. Let μ^1 be the initial derivative of η^3 along η^1; the existence and uniqueness of μ^1 follow from Lemma 8.1.1(i) (Limit Path) and Lemma 6.1.1. If $\xi^1 \subset \mu^1$, then by Lemma 6.1.1 and Remark

2.4.2(ii), no action will be taken thereafter to declare axioms for this Θ. Fix the nodes $\{\mu_i^1, \pi_i^1 : i \leq k\}$ specified in Lemma 6.1.3. As numbers enter A only for validated action, and such action has already been taken for the nodes in (π_i^1, μ_{i+1}^1), there must be an i such that $\xi^1 \in [\mu_i^1, \pi_i^1]$. Now neither μ_i^1 nor π_i^1 act to put numbers into A, so we must have $\xi^1 \in (\mu_i^1, \pi_i^1)$. Furthermore $[\mu_i^1, \pi_i^1]$ is a primary η^1-link or has cardinality 1, so π_i^1 must be switched by some node of T^0 extending the node at which the axiom is declared, but preceding the node at which validated action for ξ^1 is taken. If $\widetilde{u}$ is the use associated with π_i^1, then we will require that validation of π_i^1 occurs only if there is a change to $U \upharpoonright \widetilde{u} + 1$ between its initial derivative and the node for which π_i^1 becomes validated. But $\widetilde{u} \leq u$, so the axiom for $\Phi(A, U)$ observed through the falsity of the directing sentence for η^1 will be injured before ξ^1 places a number into A, preserving our ability to match the use of Θ to the use of Φ, and so ensure the compatibility of shared action for Θ. The above description can easily be formalized into a proof of the following lemma.

LEMMA 6.1.4. *Suppose that we are given a Δ_3 construction for which we are asked to satisfy the requirement $\Phi(A, U) = V \rightarrow \Theta(U) = V$ assigned to $\eta^3 \in T^3$. Assume that declaration of axioms for Θ obeys clause* (*i*) *of Definition* 3.1.7 *and that such declaration occurs as activated action, that the placement of numbers into A obeys clause* (*ii*) *of Definition* 3.1.7, *that the Θ use for a given argument is the same as the Φ use for the same argument, that derivatives of η^3 on T^1 which have already declared an axiom are validated only if U changes below the use associated with that derivative between the time the axiom is declared and the time of validation, and that no new axiom for Θ is declared at η^1 unless its oracle is compatible with the oracles used for axioms at $\xi^1 \subset \eta^1$ such that* up$(\xi^1) = \eta^3$ *and ξ^1 is activated along η^1. Then shared functional action for Θ is compatible.* ⊣

6.1.3. The Correction Requirements. The requirements designed to ensure that $U \not\leq A$ will be assigned to nodes $\eta^3 \in T^3$ and will have level 3. As Player II does not control U, help is required in order to satisfy this condition; we do so by establishing requirements specifying that if $\Phi(U) = A$, then we can define a computable partial functional Ξ such that $\Xi(A, U) = V$. Thus we begin defining axioms of the form $\Xi(A, U; x) = V(x)$, with A-use $\widehat{u}$ equal to the weight of the node of T^1 responsible for defining the axiom, and U-use agreeing with the U-use of the computation of $A \upharpoonright \widehat{u} + 1$ by Φ. When the number x enters V, we correct the axiom $\Xi(A, U; x) = V(x)$ by putting $\widehat{u}$ into A. The declaration of axioms for Ξ and the placement of numbers into A will be governed by Definition 6.1.5 below as well as clause (ii) of Definition 3.1.7. Thus we need only ensure that before new axioms for Ξ are declared, old axioms requiring correction are first corrected. This will be achieved through backtracking. This discussion is easily formalized to prove Lemma 6.1.6; we leave the formal proof to the reader. We first introduce Definition 6.1.5.

DEFINITION 6.1.5. Suppose that $\eta \subset \nu \subseteq \rho \in T^0$, $\nu^- = \eta$, $\eta_i \subset \rho_i$ for $i \le 1$, and $\rho_0 \subset \rho_1$. The Δ_3-*parameter specification* requires the specification of parameters for sentence decomposition to have the following properties:

(i) If an axiom $\Delta^t(A^s \restriction v, U^s \restriction u; x) \downarrow = V^s(x)$ is newly declared as action for η supported at ρ, then:
  (a) $u \le \text{wt}(\text{up}(\eta))$ and $v = \text{wt}(\text{up}(\eta)) + 1$.
  (b) $s = \text{wt}(\eta)$.
  (c) $t = \text{wt}(\nu)$.
  (d) η has Σ outcome (so is activated) along ρ.
  (e) Activated action for η at ρ imposes restraint on $A^{\text{wt}(\nu)} \restriction v$.

(ii) If an axiom $\Delta^t(A^s \restriction v, U^s \restriction u; x) \downarrow = V^s(x)$ can be newly declared as activated action for η, $A \ne \emptyset$, and η is validated along ν, then validated action for η along ν specifies that $\text{wt}(\text{up}(\eta)) \in A^{\text{wt}(\nu)}$.

LEMMA 6.1.6. *Suppose that we are given a Δ_3 construction for which we are asked to satisfy the requirement $\Phi(U) = A \to \Xi(A, U) = V$ assigned to $\eta^3 \in T^3$. Assume that declaration of axioms for Ξ and the placement of numbers into A obeys the conditions of Definition* 6.1.5 *as well as clause* (ii) *of Definition* 3.1.7. *Assume also that any derivative η^1 of η^3 on T^1 for which action has already declared an axiom is validated only if either U changes below the use associated with η^1 between the node at which the axiom is declared and the node at which η^1 is validated, or V changes causing an axiom declared as action for η^1 to become incorrect, and that no new axiom for Ξ is declared at a node σ^1 unless its oracle is compatible with the oracles for axioms at $\xi^1 \subset \sigma^1$ such that* $\text{up}(\xi^1) = \eta^3$ *and ξ^1 is activated along σ^1 and all prior axioms are correct. Then shared functional action for Ξ is compatible.* ⊣

6.1.4. Bounding A above by V. The last requirement is that $A \le_T U \oplus V$. We will not place this requirement on our tree of strategies, but rather show that properties of the construction ensure its satisfaction, within the proof. The idea here is that if V changes thereby making a Ξ axiom incorrect, we need to effectively determine whether we will place a number into A to correct this axiom. A U oracle will tell us whether the construction encounters the node of T^3 to which the requirement is assigned after the V change. If not, then there will be no A correction; if so, then we will backtrack (switching only primary links on T^2) and correct the axiom. Thus the entry of numbers into A can be determined by a $U \oplus V$ oracle.

Permitting will be a bit more complex in the context of the density theorem, and will be discussed as part of the proof of the theorem. In order to show that it is successful, we need the facts supplied by the next theorem. It is proved as Lemma 8.7.6.

LEMMA 6.1.7. *Let a Δ_3 construction be given. Suppose that $\mu^1 \subset \eta^1 \subset \pi^1 \subset \rho^1 \in T^1$, $[\mu^1, \pi^1]$ is a primary ρ^1-link, and η^1 has Π outcome along*

π^1. *Then* $\mathrm{wt}(\mathrm{up}(\pi^1)) < \mathrm{wt}(\mathrm{up}(\eta^1))$. *If, furthermore,* $\rho^1 \subset \Lambda^1 \in [T^1]$, *then* $\mathrm{up}^3(\eta^1) \not\subset \lambda^3(\Lambda^1)$.

6.2. The Density Theorem

The Density Theorem was the crowning achievement of Sacks's [22] use of the infinite injury method. The theorem states that the ordering of $\mathcal{R}$ is dense, and gave rise to hopes that the ordering was a nice one. Shoenfield [24] formulated this hope into a conjecture that was refuted by the construction of a minimal pair of computably enumerable degrees by Lachlan [7] and Yates [27]. It has since been shown that this ordering is complex.

The construction used to prove the Density Theorem is a Δ_3 construction. It utilizes both the permitting and backtracking techniques.

THEOREM 6.2.1 (The Sacks Density Theorem). *For any pair of computably enumerable degrees* $\boldsymbol{u} \leq \boldsymbol{v}$, *either* $\boldsymbol{u} = \boldsymbol{v}$, *or there is a computably enumerable degree* $\boldsymbol{a}$ *such that* $\boldsymbol{u} < \boldsymbol{a} < \boldsymbol{v}$.

PROOF. Let U and V be computably enumerable sets of degree $\boldsymbol{u}$ and $\boldsymbol{v}$ respectively. Note that $U \oplus V \equiv_T V$. We will construct a computably enumerable set A and a computable partial functional Γ, to satisfy the following requirements for all computable partial functionals Φ and Ψ:

$$R : A = \Gamma(U, V).$$

$$N_\Phi : \Phi(A, U) = V \Rightarrow \exists\Theta(\Theta(U) = V).$$

$$P_\Psi : \Psi(U) = A \Rightarrow \exists\Xi(\Xi(A, U) = V).$$

We gather these requirements, with the exception of R, into an effective list $\{R_i : i \in \mathbb{N}\}$. The theorem will follow from the satisfaction of all requirements if we let $\boldsymbol{a}$ be the degree of $U \oplus A$.

6.2.1. The R Requirement. We will use a permitting technique to construct A, and will show that this requirement is satisfied as part of our proof. This requirement is not placed on the trees of strategies.

6.2.2. The Basic N_Φ-Module. This module has a single non-terminal node α of level 3, with immediate successor α_0 (α_1, resp.) along which α is activated (validated, resp.) and has Π outcome (Σ outcome, resp.). Clauses (i)–(iii) of Definition 2.3.1 follow easily. α is responsible for defining a computable partial functional Θ_α. While the sentence $\Phi(A, U) = V$ looks like sentences we have previously encountered at level 2, the previous sentences required that we restrain an oracle under construction in order to preserve a computation. In this case, we have no control over U, so cannot preserve computations from oracle U except indirectly by allowing a node to switch only when U changes on a suitably small number. This forces us to increase the level of the requirement, thereby providing an opportunity to revise action when a new number enters the oracle. As the statement of the condition $\Phi(A, U) = V$ is Π_3, the framework requires us to take the negation of this condition as our directing sentence. The declaration of axioms will occur along the activated outcome with parameters $\eta^1_{\eta^2}$ and u_{η^2} specified by the falsity of the directing sentence for η^2; the values of these parameters may change as the construction progresses, but they will have a limiting value. No action will be necessary along the validated outcome.

Directing sentence S_α:

$$\exists \eta^2 \in \widetilde{V}_\alpha \exists x \leq \mathrm{wt}(\eta^2) \forall u \forall \eta^1 \in \widetilde{V}_{\eta^2} \exists t > \mathrm{wt}(\eta^1)(\Phi(A \restriction \mathrm{wt}(\eta^1) + 1,$$
$$U \restriction u; x)[\mathrm{wt}(\eta^1)] = V^{\mathrm{wt}(\eta^1)}(x) \rightarrow U^t \restriction u \neq U^{\mathrm{wt}(\eta^1)} \restriction u).$$

Activated action for α at $\rho \supseteq \alpha_0$:

$$\forall \eta^2 \in \widetilde{V}_\alpha \forall x \leq \mathrm{wt}(\eta^2) \exists r \forall p \geq \mathrm{wt}(\eta^1_{\eta^2})$$
$$\Theta^r_\alpha(U^{\mathrm{wt}(\eta^1_{\eta^2})} \restriction \mathrm{wt}(\eta^1_{\eta^2}); x) = \Phi(A \restriction \mathrm{wt}(\eta^1_{\eta^2}) + 1, U \restriction u_{\eta^2}; x)[\mathrm{wt}(\eta^1_{\eta^2})] \;\&$$
$$A^p \restriction \mathrm{wt}(\eta^1_{\eta^2}) + 1 = A^{\mathrm{wt}(\eta^1_{\eta^2})} \restriction \mathrm{wt}(\eta^1_{\eta^2}) + 1).$$

If S_α is true, then by Lemmas 3.1.8(i) (Σ_1 Analysis) and 4.4.6, there will be an x for which either $\Phi(A, U; x) \uparrow$ or $\Phi(A, U; x) \downarrow \neq V(x)$. Suppose that S_α is false and activated action is followed for α. Then by Lemmas 3.1.8(i) (Σ_1 Analysis) and 4.4.6, every η^2 will have a corresponding u_{η^2} and $\eta^1_{\eta^2}$, and there will be infinitely many such nodes η^2. Fix η^2, and the corresponding u_{η^2} and $\eta^1_{\eta^2}$. Then by the restraint imposed on A by activated action and the persistence of U through the falsity of the directing sentence, we have that for every $x \leq \mathrm{wt}(\eta^2)$, there is an r such that,

$$\begin{aligned} V(x) &= \Phi(A \restriction \mathrm{wt}(\eta^1_{\eta^2}) + 1, U \restriction u_{\eta^2}; x) \\ &= \Phi(A \restriction \mathrm{wt}(\eta^1_{\eta^2}) + 1, U \restriction u_{\eta^2}; x)[\mathrm{wt}(\eta^1_{\eta^2})] \\ &= \Theta^r_\alpha(U^{\mathrm{wt}(\eta^1_{\eta^2})} \restriction \mathrm{wt}(\eta^1_{\eta^2}); x) = \Theta_\alpha(U; x). \end{aligned}$$

Hence we see that this requirement is satisfied.

6.2.3. The Basic P_Ψ-Module. This module has a single non-terminal level 3 node β, with terminal immediate successor β_0 (β_1, resp.) along which β is activated (validated, resp.) and has Π (Σ outcome, resp.). Clauses (i)–(iii) of Definition 2.3.1 follow easily. The parameters $\eta^1_{\eta^2}$ and u_{η^2} obtained from the falsity of S_β can be used in the description of activated action for β, and the parameter η^2 obtained from the truth of S_β can be used in the description of validated action for β; the values of these parameters may change as the construction progresses, but they will have a limiting value.

Directing sentence S_β:

$$\exists\eta^2 \in \widetilde{V}_\beta \exists x \leq \text{wt}(\eta^2)\forall u \forall \eta^1 \in \widetilde{V}_{\eta^2}\exists t > \text{wt}(\eta^1)$$
$$(\Psi(U \restriction u; x)[\text{wt}(\eta^1)] = A^{\text{wt}(\eta^1)}(x) \to (U^t \restriction u \neq U^{\text{wt}(\eta^1)} \restriction u \vee$$
$$V^t \restriction \text{wt}(\eta^2) + 1 \neq V^{\text{wt}(\eta^1)} \restriction \text{wt}(\eta^2) + 1)).$$

Activated action for β at $\rho \supseteq \beta_0$:

$$\forall \eta^2 \in \widetilde{V}_\beta \forall x \leq \text{wt}(\eta^2)\exists r \forall q \geq \text{wt}(\eta^1_{\eta^2})$$
$$(\Xi^r_\beta(A^{\text{wt}(\eta^1_{\eta^2})} \restriction \text{wt}(\eta^1_{\eta^2}) + 1, U^{\text{wt}(\eta^1_{\eta^2})} \restriction \text{wt}(\eta^1_{\eta^2}); x) = V^{\text{wt}(\eta^1_{\eta^2})}(x)\ \&$$
$$A^q \restriction \text{wt}(\eta^1_{\eta^2}) + 1 = A^{\text{wt}(\eta^1_{\eta^2})} \restriction \text{wt}(\eta^1_{\eta^2}) + 1).$$

Validated action for β at $\rho \supseteq \beta_1$: $\forall \eta^1 \in \widetilde{V}_{\eta^2}\exists r(\text{wt}(\eta^1) \in A^r)$.

If the true path passes through β_1, then S_β is true; so as $\text{wt}(\eta^2)$ is finite and as, by Lemmas 3.1.8(i) (Σ_1 Analysis) and 4.4.6, η^2 has infinitely many derivatives along Λ^1, we must have $\Psi(U) \neq A$. If the true path passes through β_0, then S_β is false and activated action is followed for β. By Lemmas 3.1.8(i) (Σ_1 Analysis) and 4.4.6, β has infinitely many derivatives along Λ^2, so $\Psi(U) = A$ and $\Xi_\beta(A, U) = V$. We now see that in all cases, the requirement P_Ψ will be satisfied.

As we have a Δ_3 construction, the standard assignment of requirements to nodes of lower level trees will minimize conflicts between requirements. Restraint imposed for N_Φ will never conflict with positive action required for P_Ψ except for nodes whose action is governed by Lemma 3.1.9 (Restraint); injury to the directing sentence for P_Ψ will only be caused by the entry of a number into U or V that will, in turn, allow the correction of axioms for the functional $\Xi_\beta(A, U)$ associated with N_Φ. The major conflict between requirements will be the coordination of permission; P_Ψ may dictate the entry of a number x into A through action for $\eta^1 \in T^1$ at a time when the entry of x into A is permitted by V (i.e., $\Gamma(V, U; x)$ is currently undefined) but η^1 is not free, so no action for η^1 can be taken. We would then be forced to redefine $\Gamma(V, U; x) = 0$. If, at a later stage, η^1 becomes free and so able to dictate action, the permission for x to enter A may no longer be present. Thus if the dictated action is carried out, we will cause $\Gamma(V, U) \neq A$. We resolve this situation by implementing the backtracking procedure to switch links on T^2, and by placing implicit restraint on U in directing sentences.

6.2.4. *T^3*-Analysis. The standard initial assignment is followed. Clause (i) of Definition 2.7.2 follows from Lemma 2.7.4 (Standard Initial Assignment).

The standard initial sentence specification for $\eta^3 \in T^3$ is followed, with action supported at $\{\rho^3 \supset \eta^3 : \eta^3 \text{ is } \rho^3\text{-free}\}$. Clause (i) of Definition 2.10.1

will follow from the analysis for basic modules and the proof of the satisfaction of the requirement R.

6.2.5. T^2-Analysis, T^1-Analysis. The standard derived assignment is followed. Clauses (i)–(iii) of Definition 2.7.7 follow from Lemma 2.12.2 (Standard Derived Assignment).

Sentence decomposition for $\eta^2 \in T^2$ will specify that $x \leq \mathrm{wt}(\eta^2)$. Sentence decomposition for $\eta^1 \in T^1$ will specify that $u \leq \mathrm{wt}(\eta^1)$. Monotonicity of sentence decomposition follows from (2.8.1). Action for $\eta^i \in T^i$ is supported at $\{\rho^i \supset \eta^i : \eta^i \text{ is } \rho^i\text{-free and } \rho^i \text{ is not } \rho_i\text{-constrained}\}$, for $i \in \{1, 2\}$. By Lemma 4.4.7, the construction has infinite support on T^2 and T^1.

6.2.6. T^0-Analysis. The standard derived assignment is followed. Clauses (i)–(iii) of Definition 2.7.7 follow from Lemma 2.12.2 (Standard Derived Assignment).

Fix $\eta = \nu^- \subset \nu \subseteq \rho \in T^0$. Sentence decomposition for η along ρ specifies that $t \leq \mathrm{wt}(\eta)$, $r \leq \mathrm{wt}(\nu)$, and $p, q \leq \mathrm{wt}(\rho)$. Monotonicity of sentence decomposition follows from (2.8.1). (Technically, we should be bounding r on T^1. If ν^1 is the immediate successor of η^1 along which η^1 is activated, then the bound for r on T^1 would be $\mathrm{wt}(\mathrm{out}(\nu^1))$; but $\mathrm{wt}(\mathrm{out}(\nu^1)) = \nu$, so the bound given for r is the bound that would naturally occur. We have chosen to bound r on T^1 in order to more easily see that clause (ii) of Definition 3.1.7 is satisfied.) Action for η is supported at $\{\rho \supset \eta : \eta \text{ is } \rho\text{-free and } \rho \text{ is not } \rho\text{-constrained}\}$. By Lemma 4.4.7, the construction has infinite support on T^0.

6.2.7. The Construction. We say that η *requires extension* for $\xi^1 \in T^1$ if, $\xi^1 \subset \lambda(\eta)$ is a β-node, $\mathrm{up}^3(\eta) = \mathrm{up}^3(\xi^1)$, ξ^1 is activated along $\lambda(\eta)$, and the directing sentence for ξ^1 would be true at η if η were a derivative of ξ^1.

Suppose that $\eta \subset \Lambda^0$ is given. If there is a $\xi^1 \subset \lambda(\eta)$ such that η requires extension for ξ^1, fix the shortest such ξ^1, the initial derivative ξ of ξ^1 along η and let κ be the backtracking extender of η for ξ^1 from ξ (Lemma 4.4.6 (Backtracking) ensures the existence of κ); we specify that $\kappa \subset \Lambda^0$. Otherwise, the construction acts in accordance with condition (iva) of Definition 2.10.1 at η.

6.2.8. The Verification. Fix the true path $\Lambda^0 \in [T^0]$ determined by the construction, and let $\Lambda^{k+1} = \lambda(\Lambda^k)$ for $k \leq 2$. It is easily seen that the construction admits consistent action and that backtracking is normal. By Lemma 4.4.6, the construction is faithfully executed and so by Lemma 3.1.8(ii) (Σ_1 Analysis), support propagates from T^j to T^{j+1} for $j \leq 2$. We note that clauses (ii) and (iii) of Definition 3.1.7 hold, so by Lemma 3.1.9(i) (Restraint), shared set action is compatible. Also, clauses (i) and (ii) of Definition 6.1.5 hold (note that for P_Ψ requirements, $D = U \oplus A$ and a change in V causes the A oracle to change before a new Ξ-axiom is declared). The remaining hypotheses of Lemmas 6.1.4 and 6.1.6 are easily verified, so it follows from these lemmas that shared functional action is compatible. Thus clause (iv) of Definition 2.10.1 holds.

It remains to show that $A \leq_T V$; for then the theorem will follow from Theorem 2.11.5 (Modified Framework). We proceed by induction on $x \in \mathbb{N}$. Fix x. x can be placed into A only if $x = \text{wt}(\eta^1)$ for some $\eta^1 \in T^1$, so we assume that this in the case and fix η^1. Let $\eta^2 = \text{up}(\eta^1)$ and let η be the initial derivative of η^1 along Λ^0. Using V as an oracle, we can find the shortest non-constrained node $\rho \in (\eta, \Lambda^0)$ such that for all $\tau \in [\rho, \Lambda^0)$, $V^{\text{wt}(\tau)} \upharpoonright \text{wt}(\eta^2) + 1 = V \upharpoonright \text{wt}(\eta^2) + 1$; the existence of ρ follows from Lemma 4.4.7 and induction. Let σ be the shortest non-constrained node in (ρ, Λ^0). Now if $x \notin A^{\text{wt}(\sigma)}$, then by (2.8.1), Lemma 8.3.1 (Nesting), Lemma 6.1.7 and the construction, x can enter A after stage $\text{wt}(\sigma)$ only if there is a primary $\lambda(\sigma)$-link $[\mu^1, \pi^1]$ that restrains η^1, and in addition, if $[\mu^1, \pi^1]$ is the shortest such link working for a requirement N_Φ, then there must be a $\tau \in (\sigma, \Lambda^0)$ such that $U^{\text{wt}(\tau)} \upharpoonright \text{wt}(\pi^1) + 1 \neq U^{\text{wt}(\sigma)} \upharpoonright \text{wt}(\pi^1) + 1$. The existence of such a τ can be determined using U as an oracle, as can its value if it exists. It now follows from the construction (in particular the conditions under which backtracking is implemented) and Lemma 6.1.7 that if $x \notin A^{\text{wt}(\sigma)}$, then $x \in A$ iff τ exists and x is placed into A by the time we reach the shortest non-constrained node along Λ^0 that properly extends τ. We conclude that $A \leq_T V$, completing the proof of the theorem. ⊣

6.3. Summary

In this chapter, we encountered a situation in which we were forced to correct an axiom by placing a number into one of its oracle sets, in order to react to a change in the value that the axiom was designed to compute. For N_Φ requirements, restraint was imposed on the oracle set that Player II was constructing, so the change in value was generated by a change in an oracle set being constructed by Player I. Thus it was sufficient to ensure that the axiom that Player II declared had use (for Player I's oracle set) at least as long as that of the axiom being copied. For P_Ψ requirements, the axioms declared by Player II were designed to copy a set, V, constructed by Player I; however, one of the oracle sets for the axioms, A, was constructed by Player II. Thus Player II was able to react to a change in V by changing A, and thereby correcting axioms. As the set A was to be computable from V, the change in A had to take place before new axioms for the functional Ξ computing V were declared, and this was accomplished through backtracking.

Chapter 7

Σ_3 CONSTRUCTIONS

In the early days of priority arguments, constructions using this method were classified either as *finite injury* or *infinite injury*. The finite injury constructions were those that are now classified as Σ_1 or Δ_2 constructions, and the infinite injury constructions were those initially thought of as Π_2 or Δ_3 constructions. Lachlan was the first to go beyond the Δ_3 level, and the proofs, when they first appeared seemed difficult and mysterious. One such theorem was informally called the *Monster Theorem*, and the method of construction was referred to as *monstrous injury*. The classification we are using is essentially due to Harrington, who understood the level of injury to be the oracle needed to completely determine what was happening within the construction. We are following Harrington's classification, but viewing it in a slightly different light, from the viewpoint of the starting tree needed for a standard decomposition.

We will prove just one theorem at the Σ_3 level, the existence of strong minimal pairs. This theorem was chosen in advance as a test case for the framework, and reflects the author's interests in the $\forall\exists$ decision problem for $\mathcal{R}$.

7.1. Module-Respecting Derived Assignments

When basic modules are larger than a single non-terminal node, we sometimes need a careful transmission of these modules to lower levels, ensuring that successive nodes of the module have successive derivatives, whenever possible. This will be the case for the theorem of this chapter. More motivation and the proof of the lemma of this section can be found in Section 8.6.

DEFINITION 7.1.1. Let $\eta^k, \nu^k \in T^k$ be given. We say that the pair $\langle \eta^k, \nu^k \rangle$ is *connected* if $(\nu^k)^- = \eta^k$, and we say that this pair is *module-connected* if it is connected and both nodes are antiderivatives of a connected pair of nodes lying in the same realization of some basic module (as defined in Definition 2.7.1).

DEFINITION 7.1.2. A derived assignment is said to be *module-respecting* if the following conditions hold whenever $j < k$, $\eta^j, \nu^j \in T^j$ and $\eta^k, \nu^k \in T^k$:

(i) If $j = k - 1$, $\langle \eta^k, \nu^k \rangle$ is module-connected, $\langle \eta^j, \nu^j \rangle$ is connected, and η^j is the principal derivative of η^k along ν^k, then $\mathrm{up}(\eta^j) = \eta^k$.
(ii) If $j = k - 1$, $\langle \eta^k, \nu^k \rangle$ is module-connected, $\langle \eta^j, \nu^j \rangle$ is connected, $\mathrm{lev}(\eta^k) \geq k$, $\mathrm{lev}(\nu^k) \geq k$, η^k has Π outcome along ν^k, η^j has Σ outcome along ν^j, $\mathrm{up}(\eta^j) = \eta^k$, and there is no $\xi^j \subset \eta^j$ such that $\mathrm{up}(\xi^j) = \nu^k$ and ξ^j has Π outcome along ν^j, then $\mathrm{up}(\nu^j) = \nu^k$.
(iii) If $\langle \eta^k, \nu^k \rangle$ is module-connected, $\langle \eta^j, \nu^j \rangle$ is connected, $\mathrm{lev}(\eta^j) = j + 1$, $\mathrm{lev}(\nu^k) \geq j + 1$, $\mathrm{up}^k(\nu^j) = \nu^k$, and $(\mathrm{out}^{j+1}(\nu^k))^-$ has Π outcome along $\mathrm{out}^{j+1}(\nu^k)$, then $\mathrm{up}^k(\eta^j) = \eta^k$.

The following lemma is proved as Lemma 8.6.5.

LEMMA 7.1.3 (Module-Respecting Derived Assignment Lemma). *If all derived assignments on T^i for $i > j$ are module-respecting, then there is a module-respecting derived assignment for T^j satisfying conditions (i)–(iii) of Definition* 2.7.7.

7.2. Strong Minimal Pairs

Unlike other degree structures, the characterization of extensions of embeddings in $\mathcal{R}$ is fairly complicated. We prove an unpublished theorem of Slaman that demonstrates some of the complexity.

DEFINITION 7.2.1. A pair of computably enumerable degrees $\boldsymbol{a}$ and $\boldsymbol{b}$ is a *strong minimal pair* if $\boldsymbol{a}$ and $\boldsymbol{b}$ form a minimal pair of computably enumerable degrees, and for every non-computable degree $c \leq a$, $c \vee b \geq a$ (and so $c \vee b = a \vee b$).

The construction of a strong minimal pair is a Σ_3 construction. The proof uses lemmas that we have already proved, so will not look substantially different from proofs of previous theorems. It uses the same requirements as did the original $\boldsymbol{0}'''$ construction, except that that construction is essentially the reflection of a T^3 construction on the tree T^2, without going through the decomposition. We will try to motivate the proof in this section, and describe the relationship between the two approaches. We believe that is easier to present the proof using our framework, as there is a savings gained by appealing to theorems about the framework rather than having to reprove these facts. Conceptually, however, there is little difference between the proofs.

The first set of requirements to be satisfied consists of the requirements P_Φ stating that $B \neq \Phi(\emptyset)$, one for each computable partial functional Φ. These are the noncomputability requirements of Section 3.3, level 1 requirements requiring no restraint, and are satisfied as before. Further discussion of these requirements will not be presented.

The second set of requirements consists of the requirements Q_Ψ stating that $\Psi(B) \neq A$, one for each computable partial functional Ψ. On the surface, these look like the diagonalization requirements of Section 3.2, as Player II controls both A and B. However, interactions between these requirements and the requirements $M_{\Phi,W}$ introduced below will interfere with our ability to restrain B; in fact, correction for $M_{\Phi,W}$ may cause us to actively put the use of a computation $\Psi(B;x) = A(x)$ into B. Thus we will need to make infinitely many diagonalization attempts with different witnesses to satisfy Q_Ψ, raising the level of this requirement to level 2. The Π outcome of the first node of this requirement on T^2 will be followed by attempts to neutralize higher priority requirements (as viewed on T^3) $M_{\Phi,W}$ by showing that W is computable. We will make attempts for each such requirement, in order of priority, with the Π outcome providing an effective computation of W. We will be able to impose restraint on B only if all attempts follow the Σ outcome. If one of these attempts produces a Π outcome, then the corresponding higher priority requirement will have its outcome switched, allowing us to conclude that the node at which we were trying to satisfy Q_Ψ does not lie along the true path for the construction.

The last set of requirements, $M_{\Phi,W}$, tries to satisfy the strong minimal pair condition. Attempts to satisfy the requirement $M_{\Phi,W}$ are intertwined with attempts to satisfy the requirements Q_Ψ. The requirement $M_{\Phi,W}$ will begin by trying to determine whether or not $\Phi(A) = W$; the node checking for equality will have level 2, so will only be able to check for a specified length of agreement at a given stage. If it finds the desired agreement between $\Phi(A)$ and W, it acts to restrain A; its successor node, η^3, in the module will have level 3, and each initial derivative $\widehat{\eta}^2$ of η^3 on T^2 will try to declare axioms yielding successively longer lengths of agreement between $\Delta_{\widehat{\eta}^2}(B, W)$ and A. However, the restraint is imposed as action along a Σ outcome, so may be injured when a higher priority (as measured on T^3) requirement Q_Ψ acts to put a number into A. Thus when we again find that $\Phi(A)$ and W agree on a sufficiently long initial segment, W may or may not have changed below the use of the axiom declared by $\Delta_{\widehat{\eta}^2}$; but in any case, A will have changed. If W has changed, we can just declare a new axiom for $\Delta_{\widehat{\eta}^2}$; but if not, then as $M_{\Phi,W}$ will have higher overall priority than Q_Ψ, we must put a small number into B to correct the axiom declared for $\Delta_{\widehat{\eta}^2}$, thereby injuring our attempt to satisfy Q_Ψ. If we commit to making infinitely many attempts to satisfy Q_Ψ if needed, then only one of these needs to succeed; for if all attempts fail, we will argue that W is computable. The latter argument will need to focus on a single $\Delta_{\widehat{\eta}^2}$, and we will need to make infinitely many attempts to effectively define $W = \Gamma_{\eta^2}(\emptyset)$, one for each derivative η^2 of η^3 on T^2. Now Q_Ψ must take into account action for all requirements $M_{\Phi,W}$ of higher priority, so a node working for Q_Ψ and having the responsibility to restrain B will be on the true path only if all higher priority $M_{\Phi,W}$ requirements produce W changes for a particular attempt to satisfy Q_Ψ.

7.3. The Strong Minimal Pair Theorem

This section is devoted to the construction of a strong minimal pair.

THEOREM 7.3.1. *There is a strong minimal pair* **a** *and* **b** *of computably enumerable degrees.*

PROOF. We will construct computably enumerable sets A and B whose degrees comprise a strong minimal pair. It will suffice to satisfy the following requirements for every computable partial functional Φ:

$$P_\Phi : B \neq \Phi(\emptyset).$$

$$Q_\Psi : A \neq \Psi(B).$$

$$M_{\Phi,W} : \Phi(A) = W \Rightarrow (\exists\Delta(\Delta(B,W) = A) \vee \exists\Gamma(\Gamma(\emptyset) = W)).$$

We gather these requirements into an effective list $\{R_i : i \in \mathbb{N}\}$. The theorem will follow from the satisfaction of all requirements.

7.3.1. The Basic P_Φ-Module. These are the noncomputability modules of Section 3.3, and so will not be presented again. The requirements are assigned to α-nodes.

7.3.2. The Basic Q_Ψ-Module. This module begins with a non-terminal node β of level 2 that checks to see if $\Psi(B)$ computes arbitrarily long initial segments of A along the true path, and if an opportunity to diagonalize presents itself. The falsity of this sentence will produce a witness η^1 to be used in activated action, and will generate activated action that will specify that $\text{wt}(\eta^1) \notin A$, and lead to a terminal immediate extension β_0 of β along which β is activated. If the sentence is true, then validated action will place $\text{wt}(\eta^1) \in A$, but will not yet restrain B, and will bring us to a non-terminal immediate extension β_1 of β of level 1 along which β is validated. The truth of the sentence will also produce witnesses u_{η^1} and $\eta^0_{\eta^1}$ for each $\eta^1 \in V_\beta$ to be used in the sentences for β_1 The parameter β is interpreted on T^2 but we will also need to refer to its antiderivative, $\widetilde{\beta}$, on T^3. Formally, the sentences are as follows:

Directing sentence S_β:

$$\forall\eta^1 \in V_\beta \exists u \exists\eta^0 \in V_{\eta^1} \forall y < \text{wt}(\eta^1)(\Phi(B \restriction u; y)[\text{wt}(\eta^0)] \downarrow = A^{\text{wt}(\eta^0)}(y)\ \& \\ \Phi(B \restriction u; \text{wt}(\eta^1))[\text{wt}(\eta^0)] \downarrow = 0).$$

Activated action for β at $\rho \supseteq \beta_0$: $\forall t(\text{wt}(\eta^1) \notin A^t)$.

Validated action for β at $\rho \supseteq \beta_1$: $\forall\eta^1 \in V_\beta \exists r(\text{wt}(\eta^1) \in A^r)$.

The node β_1 will not act immediately, but will give certain level 3 requirements $M_{\Phi,W}$ assigned to nodes $\subset \widetilde{\beta}$ the opportunity to place $\mathrm{wt}(\eta^1)$ into B if they need to do so for correction purposes. If this opportunity is always used by a requirement $M_{\Phi,W}$, then $\widetilde{\beta}$ will not lie on the true path through T^3. But if there is an $\eta^1 \in V_\beta$ for which no $M_{\Phi,W}$ needs to take advantage of the opportunity, then we will reach β_1. The directing sentence S_{β_1} is a tautology, so only validated action can be followed, and the parameters u_{η^1} and $\eta^0_{\eta^1}$ are chosen for this instance η^1. Validated action restrains B, and brings us to a terminal immediate extension β_{11} of β_1. Formally, the sentences are as follows:

Directing sentence S_{β_1}: $0 = 0$.

Validated action for β_1 at $\rho \supseteq \beta_{11}$: $\forall t \geq \mathrm{wt}(\eta^0_{\eta^1})(B^t \restriction u_{\eta^1} = B^{\mathrm{wt}(\eta^0_{\eta^1})} \restriction u_{\eta^1})$.

If the true path $\Lambda^2 \in [T^2]$ passes through β_0 and β_0 is Λ^2-free, then either $\Psi(B)$ is not total, or there is an $x < \mathrm{wt}(\eta^1)$ such that $\Psi(B;x)\downarrow\neq A(x)$, or $\Psi(B;\mathrm{wt}(\eta^1))\downarrow\neq 0 = A(\mathrm{wt}(\eta^1))$ for some derivative η^1 of β lying on the true path through T^1. Otherwise, if $\widetilde{\beta} \subset \Lambda^3$, then no nodes will be interpolated between $\widetilde{\beta}$ and $\mathrm{up}(\beta_1)$, so $\mathrm{up}(\beta_{11}) \subset \Lambda^3$ and $\Psi(B;\mathrm{wt}(\eta^1))\downarrow = 0 \neq 1 = A(\mathrm{wt}(\eta^1))$. In all cases, we see that the requirement Q_Ψ is satisfied.

7.3.3. The Basic $M_{\Phi,W}$-Module. This module begins with a non-terminal node γ (to be interpreted on T^3) of level 2 that asks if $\Phi(A)$ and W agree on progressively longer initial segments. The node γ has terminal successor γ_0 along which it is activated and has Π outcome; in this case, the directing sentence is false and the requirement is automatically satisfied, so no activated action is necessary. The node γ has a non-terminal level 3 successor γ_1 of γ along which γ is validated and has Σ outcome; in this case, the directing sentence is true and validated action restrains A. While the parameter γ is interpreted on T^3, the sentences will have to refer to initial derivatives of γ on T^2. As $\mathrm{lev}(\gamma) = 2$, it follows from (2.6.3) that γ will have at most one derivative along any path through T^2; we let $\widehat{\gamma}$ denote an initial derivative of γ on T^2. The truth of the directing sentence for each $\eta^1 \in V_{\widehat{\gamma}}$ provides witnesses u_{η^1} and $\eta^0_{\eta^1}$ that can be used in the validated action sentence for $\widehat{\gamma}$. Formally, the sentences are:

Directing sentence $S_{\widehat{\gamma}}$:

$$\forall \eta^1 \in V_{\widehat{\gamma}} \exists u \exists \eta^0 \in V_{\eta^1} \forall x \leq \mathrm{wt}(\eta^1)(\Phi(A \restriction u; x)[\mathrm{wt}(\eta^0)] = W^{\mathrm{wt}(\eta^0)}(x)).$$

Validated action for γ along $\rho \supseteq \gamma^1$:

$$\forall \eta^1 \in V_\beta \forall t \geq \mathrm{wt}(\eta^0_{\eta^1})(A^t \restriction u_{\eta^1} = A^{\mathrm{wt}(\eta^0_{\eta^1})} \restriction u_{\eta^1}).$$

Both immediate successors γ_{10} and γ_{11} of γ_1 will be terminal; γ_1 will be activated and have Π outcome along γ_{10}, and will be validated and have Σ outcome along γ_{11}. Each initial derivative $\widehat{\gamma_1}$ of γ_1 on T^2 is responsible for defining a functional $\Delta_{\widehat{\gamma_1}}$ that computes A from $B \oplus W$, and each derivative $\eta^2 \in T^2$ of γ_1 will have an associated functional Γ_{η^2} that tries to effectively compute W, and works in conjunction with the unique $\widehat{\gamma_1} \subset \eta^2$. The action for these functionals will be taken along the activated and validated outcomes of γ_1, respectively. The directing sentence will search for an $\eta^2 \in V_{\gamma_1}$ for which $\Gamma_{\eta^2}(\emptyset)$ and W are always compatible. The existence of such an η^2 will allow $\Gamma_{\eta^2}(\emptyset)$ to provide an effective computation of W. If no such η^2 is found, then every η^2 will eventually have a derivative η^1 for which $\Delta_{\widehat{\gamma_1}}(B, W)$ does not require a B-correction to resolve an incompatibility with A, so will allow the next requirement Q_Ψ to impose B-restraint. As we always want to maintain the compatibility of $\Delta_{\widehat{\gamma_1}}(B, W)$ with A and will need the use of the computations $\Delta_{\widehat{\gamma_1}}(B, W; x)$ to be large enough to allow any W changes affecting the construction to correct $\Delta_{\widehat{\gamma_1}}$ axioms, validated action will include placing the use of a sufficiently small number (provided as the weight of a β-node) into B in order to correct the axiom. The truth of the directing sentence provides a parameter η^2 that can be used in validated action. The node $\widehat{\gamma_1}$ will be the initial derivative of γ_1 along the current path through T^2. In addition, validated action involves a parameter $p(\eta^1)$ (the weight of the β-node mentioned above) which will be defined when we bound quantifiers on T^1.

Directing sentence S_{γ_1}:

$$\exists \eta^2 \in V_{\gamma_1} \forall x \forall s (\Gamma_{\eta^2}(\emptyset; x)[s] \downarrow \rightarrow \Gamma_{\eta^2}(\emptyset; x)[s] = W^s(x)).$$

Activated action for γ_1 along $\rho \supseteq \gamma_{10}$:

$$\forall y \exists v \exists r \forall t \geq r (\Delta_{\widehat{\gamma_1}}(B \restriction v, W \restriction v; y)[t] \downarrow = A^t(y)).$$

Validated action for γ_1 along $\rho \supseteq \gamma_{11}$:

$$\forall x \exists r \forall t \geq r (\Gamma_{\eta_2}(\emptyset; x)[t] \downarrow = W^t(x) \; \& \; p(\eta^1) \in B^r).$$

If S_γ is false, then the requirement is satisfied. If S_γ is true, and the true path extends γ_{10}, then $\Delta_{\widehat{\gamma_1}}(B, W) = A$, and if the true path extends γ_{11}, then there is an η^2 such that $\Gamma_{\eta^2}(\emptyset) = W$. The requirement $M_{\Phi,W}$ is satisfied in all cases.

7.3.4. T^3-Analysis. The standard initial assignment process is followed, so clause (i) of Definition 2.7.2 follows from Lemma 2.7.4 (Standard Initial Assignment). Hence it suffices to show that all requirements are satisfied.

The standard initial sentence specification for $\eta^3 \in T^3$ is followed, with action supported at $\{\rho^3 \supset \eta^3 : \eta^3 \text{ is } \rho^3\text{-free}\}$. Clause (i) of Definition 2.10.1 follows from the analysis of basic modules.

7.3.5. T^2-Analysis. We follow a modified derived assignment defined as follows for successors of the Π outcomes of β-nodes $\eta^2 \in T^2$. For each such η^2, let

$$X_{\eta^2} = \{\xi^3 \subset \mathrm{up}(\eta^2) : \xi^3 \text{ is a } \gamma_1\text{-node having } \Pi \text{ outcome along } \mathrm{up}(\eta^2)\}.$$

We assign a derivative of the shortest element of X_{η^2} to the Π outcome of η^2, and continue to assign derivatives of nodes in X_{η^2} without repetition, in order of length, until we either follow a Π outcome of one of these derivatives, or have assigned a derivative of each member of X_{η^2} along the given interval of T^2. In the latter case, we end by assigning the corresponding β_1-node to the last Σ outcome. We refer to the above description of a β-node followed by γ_1-nodes and perhaps a final β_1-node as a basic module. In all other cases, we follow the instructions for the standard derived assignment. Clauses (i)–(iii) of Definition 2.7.7 are easily verified.

Sentence decomposition is not necessary on T^2 as the witnesses are provided in the sentence. Monotonicity of sentence decomposition follows from (2.8.1). Action for $\eta^i \in T^2$ is supported at $\{\rho^2 \supset \eta^2 : \eta^2 \text{ is } \rho^2\text{-free}\}$. By Lemma 3.1.8(i) ($\Sigma_1$ Analysis), the construction has infinite support on T^2, and support propagates from T^2 to T^3.

7.3.6. T^1-Analysis. We again follow a modified derived assignment that reflects and expands on the derived assignment on T^2. On T^2, we have created modules that begin with a β-node, follow its Π outcome with γ_1-nodes as long as they have Σ outcome, and when the last such node is placed along a path and has Σ outcome, we follow it with the β_1-node corresponding to the starting β-node. We want what is essentially a module-respecting image on T^1, i.e., that the choice of antiderivative give preference to derivatives of the γ_1-nodes and β_1-nodes inserted, as described above, after a β-node. However, each γ_1 node in this module on T^2 has a unique corresponding shorter γ-node (recall that the γ-nodes are of level 2) and follows its Π outcome. We require this γ-node to supply a derivative immediately preceding the corresponding γ_1-node. This description is a modification of the module-respecting derived assignment, and by the proof of Lemma 7.1.3 (presented for Lemma 8.6.5), is seen to satisfy clauses (i)–(iii) of Definition 2.7.7. If a γ_1-node $\eta^1 \in T^1$ is part of such a module created for the β-node $\xi^1 \in T^1$, we define $p(\eta^1) = \xi^1$; if no such β-node exists, then $p(\eta^1)$ is undefined, and action referring to $p(\eta^1)$ is not implemented.

Sentence decomposition is not necessary for α-nodes until we reach T^0, and is already provided for β-nodes, β_1-nodes and γ-nodes. For γ_1-nodes, we bound y by the weight of the initial derivative of $\mathrm{up}(\eta^1)$ along η^1, x is bounded by $\mathrm{wt}((\eta^1)^-)$, and s is bounded by $\mathrm{wt}((\mathrm{out}(\eta^1))^-)$. Monotonicity of sentence decomposition follows from (2.8.1). Action for $\eta^1 \in T^1$ is supported at

$\{\rho^1 \supset \eta^1 : \eta^1 \text{ is } \rho^1\text{-free}\}$. By Lemma 3.1.8(i) ($\Sigma_1$ Analysis), the construction has infinite support on T^1, and support propagates from T^1 to T^2.

7.3.7. T^0-Analysis. The module-respecting derived assignment is followed (we treat the block of γ, γ_1 and β_1 nodes following a β-node on T^1 as described above as a module). Clauses (i)–(iii) of Definition 2.7.7 follow from Lemma 7.1.3.

Fix $\eta = \nu^- \subset \nu \subseteq \rho \in T^0$. Sentence decomposition for all nodes is as follows. The bounds for all sentences are: $u \leq \text{wt}(\lambda(\eta))$, $v = \text{wt}(\text{up}(\eta^-))$, $t = \text{wt}(\rho)$, and $r = \text{wt}(\nu)$. Monotonicity of sentence decomposition follows from (2.8.1) and (2.8.4). Action is supported at $\{\rho \supset \eta : \eta \text{ is } \rho\text{-free}\}$. By Lemma 3.1.8(i) ($\Sigma_1$ Analysis), the construction has infinite support on T^1, and support propagates from T^0 to T^1.

7.3.8. The Construction. The construction is faithfully executed.

7.3.9. The Verification. Fix the true path $\Lambda^0 \in [T^0]$ determined by the construction, and let $\Lambda^{k+1} = \lambda(\Lambda^k)$ for $k \leq 2$. Conditions (i) and (ii) of Definition 3.4.1 are satisfied, so by Lemma 3.4.2 (Delayed Set Action), shared set action is compatible. It is easily seen that the construction admits consistent action.

We note that if $\eta^1 \in T^1$ is a γ_1-node that becomes validated during the course of the construction, then its directing sentence must be true when evaluated at the initial derivative η of η^1 along Λ^0, so by the module-respecting feature of the assignment procedure on T^1, η^- is the principal derivative of $(\eta^1)^-$ along Λ^0; hence $\eta^- = (\text{out}(\eta^1))^-$. Axioms for $\Gamma_{\eta^2}(\emptyset)$ are declared as action for η only when $\Gamma_{\eta^2}^{\text{wt}(\eta^-)}(\emptyset)$ and $W^{\text{wt}(\eta^-)}$ are compatible and all such axioms output values compatible with $W^{\text{wt}(\eta^-)}$. Thus shared functional axioms for $\Gamma_{\eta^2}(\emptyset)$ are compatible.

It remains to be shown that shared functional action for $\Delta_{\widehat{\eta}^2}(B, W)$ is compatible, where $\widehat{\eta}^2$ is an initial derivative of a γ_1-node η^3 along a path through T^2. Suppose that $\langle \eta_0, \rho_0 \rangle$ and $\langle \eta_1, \rho_1 \rangle$ share an axiom for $\Delta_{\widehat{\eta}^2}(B, W)$ on argument y, with $\rho_0 \subseteq \rho_1$. For $j \leq 3$ and $i \leq 1$, let $\eta_i^j = \text{up}^j(\eta_i)$; note that $\eta^3 = \eta_i^3$ and that $\widehat{\eta}^2 \subseteq \eta_i^2$ for $i \leq 1$. Let $\widehat{\eta}^1$ be the initial derivative of $\widehat{\eta}^2$ on T^1. Without loss of generality, we can assume that this axiom is not shared by any $\langle \eta, \rho \rangle$ with $\rho_0 \subset \rho \subset \rho_1$. Before proceeding with the proof, we make several observations whose verification is left to the reader.

(O1) If action for $\xi \subset \Lambda^0$ places a number into A, then ξ is a β-node obeying clause (ii) of Definition 3.1.7.

(O2) If $i \leq 1$ and $\widetilde{\eta}_i^1$ is the initial derivative of η_i^2 along η_i^1, then $y \leq \text{wt}(\widetilde{\eta}_i^1)$.

(O3) By Remark 2.4.2(ii), no node $\subset \widehat{\eta}^1$ can be switched by any node in $(\rho_0, \rho_1]$.

It follows from (O1), (O2) and (2.8.1) that the shared action will be compatible unless there is a $\tau \in (\rho_0, \rho_1]$ and a β-node $\sigma \subset \rho$ such that $\sigma^1 = \text{up}(\sigma) \subset$

$\widetilde{\eta}_0^1 \wedge \widetilde{\eta}_1^1$, σ is activated along $\widetilde{\eta}_0^1$ (and so $\mathrm{wt}(\mathrm{up}(\sigma)) \notin A^{\mathrm{wt}(\tau^-)}$) and σ is validated along $\widetilde{\eta}_1^1$ (and so $\mathrm{wt}(\mathrm{up}(\sigma)) \in A^{\mathrm{wt}(\tau)}$). We assume that such σ and τ exist, else there is nothing more to show; by clause (ii) of Definition 3.1.7, we may assume that $\tau^- = \sigma$. Note that $\mathrm{up}^3(\eta_0) = \mathrm{up}^3(\eta_1)$; we let $\eta^3 = \mathrm{up}^3(\eta^0)$.

The determination of antiderivatives now plays a crucial role. Antiderivatives are assigned giving preference to the γ_1-nodes having Π outcome along $\mathrm{up}^3(\sigma)$, and when applicable, to their corresponding γ-node. If such a γ-node $\xi \in [\tau, \rho_1)$ is the principal derivative of $\mathrm{up}^2(\xi)$ and has Σ outcome along ρ_1, or if such a γ_1-node $\xi \in [\tau, \rho_1)$ has Π outcome along ρ_1, then $\widehat{\eta}^2$ will not lie on the current path through T^2 at ξ. By (O3), if every such γ-node has Π outcome at ρ_1 and every such γ_1-node has Σ outcome at ρ_1, then we will reach a γ_1-node $\xi^1 \in T^1$ in the generalized basic module determined by σ^1 whose activated outcome, by (2.8.4), would be responsible to declare an axiom for $\Delta_{\widehat{\eta}^2}(B, W; y)$; let ξ be the first derivative of ξ^1 along Λ^0 reached in this way, and let ν be its immediate successor along Λ^0. By (O2) applied to the smallest y for which such axioms are shared, ξ will be an initial derivative, so we will begin working on a functional Γ never before encountered in the construction, and so will act to validate η_1 and thus to place $\mathrm{wt}(\sigma^1) \in B^{\mathrm{wt}(\nu)}$. By (2.8.1), (O1), (O2), and as the B-use of the axiom declared for η_0 has use $\mathrm{wt}((\eta_0^1)^-) > \mathrm{wt}(\sigma^1)$, the oracles for the axioms shared by $\langle \eta_0, \rho_0 \rangle$ and $\langle \eta_1, \rho_1 \rangle$ will be incompatible; we thus conclude that the shared axioms are compatible.

As the construction is faithfully executed, Lemma 3.1.6 now allows us to conclude that clause (iv) of Definition 2.10.1 holds. The theorem now follows from Theorem 2.11.5 (Modified Framework). ⊣

REMARK 7.3.2. We note that the proof that shared functional axioms for Δ are compatible follows for any construction using the modules described above for $M_{\Phi,W}$ if all nodes placing numbers into A satisfy clause (ii) of Definition 3.1.7 and if they impose B-restraint, then they form generalized modules together with the $M_{\Phi,W}$ requirements computing A, as described in the above construction.

7.4. Some Comments

This concludes the first part of the book in which we have presented proofs of known theorems, using the framework and appealing to theorems about the framework that are proved in the next two chapters. It is, perhaps, a good point to take stock of the guiding philosophy used to select the theorems.

We had two goals in undertaking this project; showing the general applicability of our methods for proving theorems about the computably enumerable degrees whose proofs use the priority method, and showing the usefulness of this approach for proving theorems requiring priority methods of large arithmetical complexity. In order to devise what we considered to be a fair test for the first goal, we selected the theorems whose proofs we wanted to present in advance, trying to represent a broad range of combinatorial schemes. We have faithfully adhered to that selection of theorems with one exception, Friedberg's Maximal Set Theorem. An earlier version of this monograph contained a proof of the Maximal Set Theorem; but as this is not a theorem about the computably enumerable degrees, we decided to eliminate it. We did leave in two other theorems that are not completely about the computably enumerable degrees, namely the Sacks Splitting Theorem and the d-c.e. Degree Theorem, but that either have consequences for the computably enumerable degrees or are closely related. In addition, we did later look at several other theorems, including the embedding of M_3, i.e., the 1-3-1 lattice into the c.e. degrees, the Lachlan Non-Bounding Theorem which states that there is a non-zero c.e. degree that does not bound a minimal pair, and the Lachlan Non-Splitting Theorem stating that there is an interval of c.e. degrees in which the top degree cannot be non-trivially split above the bottom degree. In all cases, we were able to convince ourselves that a proof could be presented using the framework.

The second goal was to prove theorems about arbitrarily high levels of the arithmetical hierarchy. We prove a new theorem in Chapter 10; that theorem characterizes the finite jump-posets that can be embedded into the c.e. degrees. We have also been able to convince ourselves that the framework can be used to construct incomparable arithmetical degrees, and more generally, to prove theorems about the arithmetical degrees. There is much that can be done in this area.

It is our feeling that any priority argument proof of properties of the c.e. degrees can be presented in our framework. There is no way we know of to verify this feeling; rather, the feeling is the result of the experience we have had so far working with the framework. There will be cases, such as the embedding of the 1-3-1 lattice, where the proof using the framework may seem more cumbersome; but ultimately, every proof we have considered seems to have a translation into the framework. Thus we hope that it will become a valuable tool for research

Chapter 8

PATHS AND LINKS

There are many combinatorial properties that are shared by all constructions using the trees of strategies approach, some of which are relevant only to constructions of a fixed level. We gather some of these properties together in this chapter. We have attempted to isolate many combinatorial lemmas in this chapter, so that the proofs presented in other chapters depend only on the statements of these lemmas together with some special properties of the construction.

The lemmas of the current chapter are true of all assignments of requirements and derivatives to trees that satisfy the conditions of Chapter 2. These lemmas deal, among other things, with existence of initial and principal derivatives, an analysis of the ability to take both switching and nonswitching extensions of nodes, an analysis of the link creation process, and an analysis of the situations in which we can find free derivatives of a node. We begin, in Section 8.1, with some lemmas about the path generating and outcome functions λ and out. In Section 8.2, we demonstrate the ability to take both switching and nonswitching extensions. We prove some technical lemmas about links in Section 8.3, and apply these in Section 8.4 to conclude that certain nodes will be free at various stages of the construction. More general theorems are proved in Section 8.5. The remaining sections deal with lemmas that are level-specific, i.e., depend on the specific starting tree, or are more technical in nature.

Many of the lemmas deal with relationships between successive trees T^k and T^{k+1}. We will state these lemmas in full generality, but frequently present only the proof for $k = 0$, noting that the proofs relativize. We also fix n to be the level of the construction under consideration.

8.1. Paths

The path generation process is designed to have many nice, uniform properties. We describe some of those properties in this section. We begin with a lemma that states that paths through the trees are either infinite or end at nodes of the tree to which no requirements are assigned, and that initial and principal derivatives exist. This lemma is used many times, and provides an important tool in the analysis of derived assignments. We first use this lemma to spell out nice properties of the outcome function. Finally, we analyze the effect of a switching node on the path computation process. The ability to determine the nodes on each tree that were switched by the action of taking immediate extensions on T^0 was the motivating factor in the definition of trees and of the path computation function λ.

LEMMA 8.1.1 (Limit Path Lemma). *Fix $k \in [0, n)$ and $\rho^k \in T^k \cup [T^k]$. Let $\rho^{k+1} = \lambda(\rho^k)$ if $|\rho^k| < \infty$, and let $\rho^{k+1} = \lim\{\lambda(\eta^k) : \eta^k \subset \rho^k\}$ if $|\rho^k| = \infty$. Then:*

(*i*) *If $\eta^{k+1} \subset \lambda(\rho^k)$, then η^{k+1} has an initial derivative η^k along ρ^k and $\lambda(\eta^k) = \eta^{k+1}$.*

(*ii*) *If $\eta^{k+1} \subset \lambda(\rho^k)$, then there is a $\nu^k \subseteq \rho^k$ such that $\pi^k = (\nu^k)^-$ is the principal derivative of η^{k+1} along ρ^k, $(\lambda(\nu^k))^- = \eta^{k+1}$, and for all $\xi^k \subseteq \rho^k$, $\lambda(\nu^k) \subseteq \lambda(\xi^k)$ iff $\nu^k \subseteq \xi^k$.*

(*iii*) *If $|\rho^k| = \infty$, then either $|\rho^{k+1}| = \infty$, or no $\xi^{k+1} \supseteq \rho^{k+1}$ has an antiderivative $\xi^k \subset \rho^k$.*

(*iv*) *If $|\rho^k| = \infty$, then every block begun at $\delta^k \subset \rho^k$ is completed along ρ^k.*

PROOF. We prove the lemma for $k = 0$, noting that the proof relativizes. As is our convention, we will drop the superscript 0.

(i): By Remark 2.4.2(iii) and as $\eta^1 \subset \lambda(\rho)$, η^1 must have a derivative along ρ; the shortest derivative η of η^1 along ρ is the initial derivative of η^1 along ρ. Now (2.6.1) tells us that $\lambda(\eta) \supseteq \eta^1$, and by Remark 2.4.2(iii), $\lambda(\eta) \not\supset \eta^1$. Hence $\lambda(\eta) = \eta^1$.

(ii): Suppose that $\eta^1 \subset \lambda(\rho)$. By Remark 2.4.2(iii), η^1 must have a derivative $\subset \rho^1$, hence must have a principal derivative $\pi \subset \rho$. Fix $\nu \subseteq \rho$ such that $\nu^- = \pi$. By (2.6.1), $\lambda(\pi) \supseteq \eta^1$, and so by Definition 2.4.1, $\lambda(\nu) \supseteq \eta^{1\frown}\langle\nu\rangle$. Let $\nu^1 = \eta^{1\frown}\langle\nu\rangle$. By (2.6.1) and Definition 2.4.1, if $\text{up}(\xi) = \eta^{1\frown}\langle\nu\rangle$ then $\nu \subseteq \xi$. Hence again by Definition 2.4.1, we cannot have $\lambda(\xi) \supset \nu^1$ for any $\xi \subseteq \nu$. We thus conclude that $\lambda(\nu) = \nu^1$.

As π is the principal derivative of η^1 along ρ and $\eta^1 \subset \lambda(\rho)$, we must have $\nu^1 \subseteq \lambda(\rho)$; and we have already noted that $\nu^1 = \lambda(\nu)$. It now follows from Remark 2.4.2(ii) that $\nu^1 \subseteq \lambda(\xi)$ for all $\xi \in [\nu, \rho]$. By Definition 2.4.1, if $\xi \subset \nu$ then $\lambda(\xi) \not\supseteq \nu^1$, so (ii) follows.

(iii): Suppose that $|\rho| = \infty$ and $|\rho^1| < \infty$. By (2.6.1) and Definition 2.4.1, there can be no $\sigma \subset \rho$ such that $\text{up}(\sigma) \supset \rho^1$ unless there is a $\xi \subset \rho$ such that $\text{up}(\xi) = \rho^1$. Hence it suffices to show that there is no $\xi \subset \rho$ such that $\text{up}(\xi) = \rho^1$. We assume that such a ξ exists, and obtain a contradiction.

Fix the shortest $\eta \subset \rho$ such that $\lambda(\eta) \supseteq \rho^1$. By Remark 2.4.2(ii), $\lambda(\tau) \supseteq \rho^1$ for all $\tau \in [\eta, \rho)$. As there is a $\xi \subset \rho$ such that $\text{up}(\xi) = \rho^1$, ρ^1 must have a principal derivative $\pi \subset \rho$; fix $\nu \subset \rho$ such that $\nu^- = \pi$. By (ii), $\lambda(\tau) \supseteq \rho^{1\frown}\langle\nu\rangle$

for all $\tau \in [\nu, \rho)$, contrary to our definition of $\rho^1 = \lim\{\lambda(\eta) : \eta \subset \rho\}$, so (iii) follows.

(iv): We proceed by reverse induction on $j < n$, so may assume that (iv) holds for $j = 1$ and verify (iv) for $j = 0$. (Note that if $n = 1$, then either ρ^1 is finite, or (iv) follows from clause (i) of Definition 2.3.1 and Definition 2.7.1.) Suppose that $|\rho| = \infty$ and a block is begun at $\delta \subset \rho$. By Definition 2.7.5 and (i), blocks on T^0 are completed at nodes that are initial derivatives of nodes of T^1 completing blocks, so this is only possible if $|\lambda(\rho)| < \infty$. Thus by Remark 2.4.2(ii), there is an $\eta \subset \rho$ such that for all ξ satisfying $\eta \subseteq \xi \subseteq \rho$, $\lambda(\xi) = \rho^1$. By clause (ii) of Definition 2.7.7, there is a $\tau \subset \rho$ that is a derivative of the longest node $\subset \rho^1$ having level 1 and Π outcome along ρ^1. By Definition 2.7.5, the δ-block along ρ is completed at such a τ. ⊣

The next lemma describes some useful properties of the function out.

LEMMA 8.1.2 (Out Lemma). *Fix $k \leq n$ and $\rho^k \in T^k$. Then:*

(i) If $k > 0$ and ρ^k is in the image of T^{k-1} under λ, then $\lambda(\text{out}(\rho^k)) = \rho^k$.
(ii) If $k < n$ and $|\rho^k| > 0$, then there is a unique $\rho^{k+1} \in \lambda(T^k)$ such that $\text{out}(\rho^{k+1}) = \rho^k$.

PROOF. We prove the lemma for $k = 0$, noting that the proof relativizes.

(i): $\rho = \rho^{-\frown}\langle \text{out}(\rho)\rangle$, and $(\text{out}(\rho))^-$ is the principal derivative of ρ^- along $\text{out}(\rho)$. Hence (i) follows from Lemma 8.1.1(ii) (Limit Path).

(ii): Let $\nu = \rho^-$, let $\nu^1 = \text{up}(\nu)$, and let $\rho^1 = \nu^{1\frown}\langle\rho\rangle$. Then $\text{out}(\rho^1) = \rho$. Uniqueness follows once we note that if $\text{out}(\sigma^1) = \rho$, then $\sigma^1 = ((\sigma^1)^-)^\frown\langle\rho\rangle$ and $\text{up}(\rho^-) = (\sigma^1)^-$. Hence $(\sigma^1)^- = \nu^1$ and so $\sigma^1 = \rho^1$. ⊣

We will need a precise analysis of the behavior of the path generating function λ. Suppose that $\eta^k \in T^k$ and $|\eta^k| > 0$. We discuss the relationship of the path computed by $\lambda^j((\eta^k)^-)$ to the path computed by $\lambda^j(\eta^k)$ for all $j \in [k, n]$. Three types of phenomena can occur, and induce a partition of $[k, n]$ into three (possibly empty) connected intervals.

There will be a $p \geq k$ such that for all $j \in [k, p]$, $\lambda^j(\eta^k)$ is an immediate successor of $\lambda^j((\eta^k)^-) = \text{up}^j((\eta^k)^-)$. η^k is not j-switching for any $j \in [k, p]$.

If $p \neq n$, then there will be two possibilities. The first is that $\text{up}^p((\eta^k)^-) = (\lambda^p(\eta^k))^-$ has Π outcome along $\lambda^p(\eta^k)$. Then η^k will be $(p+1)$-switching, and will switch $\text{up}^j((\eta^k)^-)$ for all $j \in [p+1, n]$. η^k will switch the outcome of $\text{up}^{p+1}((\eta^k)^-)$ from a Π outcome along $\lambda^{p+1}((\eta^k)^-)$ to a Σ outcome along $\lambda^{p+1}(\eta^k)$. (We will show in Lemma 8.4.1 (Free Extension) that, in this case, for all $j \geq p+1$, η^k will switch the outcome of $\text{up}^j((\eta^k)^-)$ from a Π outcome along $\lambda^j((\eta^k)^-)$ to a Σ outcome along $\lambda^j(\eta^k)$ if $j - (p+1)$ is even, and from a Σ outcome along $\lambda^j((\eta^k)^-)$ to a Π outcome along $\lambda^j(\eta^k)$ if $j - (p+1)$ is odd.) There will be an $s \geq p$ such that for all $j \in [p+1, s]$, $\text{up}^j((\eta^k)^-)$ will be the principal derivative of $\text{up}^{j+1}((\eta^k)^-)$ along $\lambda^j(\eta^k)$, and $\lambda^j(\eta^k)$ will be an immediate successor of $\text{up}^j((\eta^k)^-)$. $[p+1, s]$ is the interval where the second type of phenomenon occurs.

If $s < n$, then the third type of phenomenon begins at $s+1$ ($s = p$ if $\text{up}^p((\eta^k)^-) = (\lambda^p(\eta^k))^-$ has Σ outcome along $\lambda^p(\eta^k)$, that is the second possibility alluded to in the preceding paragraph, and if this is the case, then η^k is not switching). Here $\text{up}^s((\eta^k)^-)$ will have Σ outcome along $\lambda^s(\eta^k)$ and will not be the principal derivative of $\text{up}^{s+1}((\eta^k)^-)$ along $\lambda^s(\eta^k)$. Thus by Definition 2.4.1, $\lambda^{s+1}(\eta^k) = \lambda(\text{up}^s((\eta^k)^-))$, so $\lambda^t(\eta^k) = \lambda^t(\text{up}^s((\eta^k)^-))$ for all $t \in [s+1, n]$.

The three types of phenomena mentioned above can be observed if we consider the usual way to satisfy a thickness requirement on T^2. η^0 decides, for $(\eta^0)^- = \xi^0$, whether to place an additional element x into a set S that is to be either a finite subset of an infinite computable set R, or all of R. x will be placed into S exactly if ξ^0 has Π outcome along η^0. First consider the case in which ξ^0 is an initial derivative of $\xi^1 = \text{up}(\xi^0)$. If ξ^1 is an initial derivative of $\text{up}(\xi^1)$, then we will have $p = s = 2$; otherwise, we will have $p = 1$, and $s = 1$ ($s = 2$, resp.) if ξ^0 has Π outcome (Σ outcome, resp.) along η^0. In the latter cases, if ξ^0 has Π outcome along η^0, then we have again found confirmation that S should be infinite, and so place x into S; and if ξ^0 has Σ outcome along η^0, then we have reaffirmed the decision that no more numbers be placed into S. Next, consider the case in which ξ^0 is not an initial derivative of $\xi^1 = \text{up}(\xi^0)$. If ξ^0 has Σ outcome along η^0, then there is no change in path computation on T^1 or T^2, so $p = s = 0$. Otherwise, there is a change in path computation on T^1, so $p = 0$ and $s \geq 1$. If $\xi^1 = \text{up}(\xi^0)$ is not an initial derivative, then $s = 1$; otherwise, $s = 2$.

LEMMA 8.1.3 (λ-Behavior Lemma). *Fix $\nu^k \in T^k$. Then there are p and s such that $k \leq p \leq s \leq n$ and the following conditions hold*:

(*i*) *For all $i \in [k, p]$, $\lambda^i((\nu^k)^-) = \text{up}^i((\nu^k)^-) = (\lambda^i(\nu^k))^-$, if $i < p$ then $\lambda^i((\nu^k)^-)$ is the initial derivative of $\lambda^{i+1}((\nu^k)^-)$ along $\lambda^i(\nu^k)$, and if $i > k$ then* $\text{out}(\lambda^i(\nu^k)) = \lambda^{i-1}(\nu^k)$.

(*ii*) *For all $i \in (p, s]$*:
  (*a*) $\text{up}^i((\nu^k)^-) = (\lambda^i(\nu^k))^-$; *and*
  (*b*) $\lambda^i((\nu^k)^-) \mid \lambda^i(\nu^k) = (\text{up}^i((\nu^k)^-))^\frown\langle\lambda^{i-1}(\nu^k)\rangle$.

(*iii*) *For all $i \in (s, n]$*, $\lambda^i(\nu^k) = \lambda^i(\text{up}^s((\nu^k)^-))$.

PROOF. We prove the lemma for $k = 0$, noting that the proof relativizes. We verify (i)–(iii) by induction on $|\nu|$. If $i = 0$, then $\nu^- = \lambda^0(\nu^-) = (\lambda^0(\nu))^- = \text{up}^0(\nu^-)$. Let $p \leq n$ be the largest number such that $\lambda^p(\nu^-) = (\lambda^p(\nu))^- = \text{up}^p(\nu^-)$. (i) now follows from Definition 2.4.1.

First suppose that ν is nonswitching. We then set $s = p$ and note that (ii) holds vacuously, and that (iii) holds vacuously if $s = n$. So suppose that $s < n$. As it is not the case that $\lambda^{s+1}(\nu^-) = (\lambda^{s+1}(\nu))^- = \text{up}^{s+1}(\nu^-)$, it follows from Lemma 8.1.1(i) (Limit Path) that $(\lambda^s(\nu))^-$ cannot be an initial derivative of $\text{up}^{s+1}(\nu^-)$. So since ν is nonswitching, it follows from Definition 2.4.1 that $\lambda^{s+1}(\nu) = \lambda^{s+1}(\nu^-)$; hence for all $i \in [s+1, n]$,

$$\lambda^i(\nu) = \lambda^i(\lambda^{s+1}(\nu)) = \lambda^i(\lambda^{s+1}(\nu)^-) = \lambda^i(\lambda^s(\nu^-)) = \lambda^i(\text{up}^s(\nu^-)),$$

and (iii) must hold.

Suppose that ν is switching. By (i), let $\eta^p = \text{up}^p(\nu^-) = (\lambda^p(\nu))^-$, and let $\eta^{p+1} = \text{up}^{p+1}(\nu^-)$. We first show that ν is $(p+1)$-switching, and that η^p has Π outcome along $\lambda^p(\nu)$. For suppose that η^p has Σ outcome along $\lambda^p(\nu)$ in order to obtain a contradiction. It cannot be the case that η^p is an initial derivative of η^{p+1}, else the conditions of (i) would hold for $p+1$ in place of p, contrary to the choice of p. But if η^p is not an initial derivative of η^{p+1}, then as η^p has Σ outcome along $\lambda^p(\nu)$, $\lambda^{p+1}(\nu) = \lambda(\eta^p) = \lambda(\lambda^p(\nu^-))$ and so ν is nonswitching, contrary to assumption. We conclude that η^p has Π outcome along $\lambda^p(\nu)$ and is the principal derivative of η^{p+1} along ν, so by Lemma 8.1.1(ii) (Limit Path), $\lambda^{p+1}(\nu) = \eta^{p+1\frown}\langle\lambda^p(\nu)\rangle$ and ν is $(p+1)$-switching. Thus (ii) holds for $i = p+1$. Fix the least $s \in (p, n)$, if any, such that the conditions of (ii) fail for $i = s+1$; otherwise, let $s = n$. (ii) now follows.

As $s+1 > p+1$, we note that $\text{up}^s(\nu^-)$ cannot be the principal derivative of $\text{up}^{s+1}(\nu^-)$ along $\lambda^s(\nu)$, else by Lemma 8.1.1(ii) (Limit Path), the condition specified in (ii) would hold for $i = s+1$. Thus $\text{up}^s(\nu^-) = (\lambda^s(\nu))^-$ cannot have Π outcome along $\lambda^s(\nu)$, nor can $\text{up}^s(\nu^-)$ be the initial derivative of $\text{up}^{s+1}(\nu^-)$ along $\lambda^{s+1}(\nu)$. Thus if $s < n$, then by Definition 2.4.1, $\lambda^{s+1}(\nu) = \lambda((\lambda^s(\nu))^-)$, so $\lambda^i(\nu) = \lambda^i((\lambda^s(\nu))^-) = \lambda^i(\text{up}^s(\nu^-))$ for all $i \in (s, n]$. ⊣

From now on, whenever we write $\Lambda^k \in [T^k]$, we assume that there is a $\Lambda^0 \in [T^0]$ such that $\Lambda^k = \lambda^k(\Lambda^0)$. Similarly, if we write $\eta^k \in T^k$, we assume that $\eta^k \subset \Lambda^k$ for some $\Lambda^k \in [T^k]$. If this is not the case, then η^k and Λ^k are irrelevant to our construction.

8.2. Switching and Nonswitching Extensions

The λ-Behavior Lemma describes the effect of extending nodes on the path computation process. We now want to look at the inverse question, namely, can we get immediate extensions ν^k of η^k that are nonswitching extensions, and can we get switching extensions. Lemma 8.2.1 (Nonswitching Extension) provides the answer for nonswitching extensions; it shows that we can always find a nonswitching extension, and that the η^k-links and the ν^k-links coincide when we take nonswitching extensions. Lemma 8.2.2 (Switching Extension) answers the question for switching extensions, and proves that we can find a switching extension unless η^k is an initial derivative of $\text{up}^j(\eta^k)$ for all $j \in [k, n]$. In both cases, the determination of whether the extension is switching or nonswitching depends only on whether η^k has Σ outcome or Π outcome along ν^k.

LEMMA 8.2.1 (Nonswitching Extension Lemma). *Fix $k \geq 1$ and $\eta^k \in T^k$. Then either $\eta^{k\frown}\langle\beta^{k-1}\rangle$ will be nonswitching for every $\beta^{k-1} \in T^{k-1}$ such that $\eta^{k\frown}\langle\beta^{k-1}\rangle \in T^k$ and $(\beta^{k-1})^-$ has Π outcome along β^{k-1}, or $\eta^{k\frown}\langle\beta^{k-1}\rangle$ is nonswitching for every $\beta^{k-1} \in T^{k-1}$ such that $\eta^{k\frown}\langle\beta^{k-1}\rangle \in T^k$ and $(\beta^{k-1})^-$ has Σ outcome along β^{k-1}. Moreover, if $\eta^{k\frown}\langle\beta^{k-1}\rangle$ is a nonswitching extension of η^k, then for all $j \geq k$, the $\lambda^j(\eta^k)$-links and the $\lambda^j(\eta^{k\frown}\langle\beta^{k-1}\rangle)$-links coincide.*

PROOF. We prove the lemma for $k = 1$, noting that the proof relativizes. We proceed by induction on $n - k$, and present the proof with notation that relativizes straightforwardly. We note that no node on T^n is switching.

Suppose that $n > 1$ and that η^1 is not the initial derivative of $\text{up}(\eta^1)$ along η^1. Fix $\beta \in T^0$ such that β^- has Σ outcome along β. By Definition 2.4.1, $\lambda(\eta^{1\frown}\langle\beta\rangle) = \lambda(\eta^1)$. Thus for all $j \in [2, n]$,

$$\lambda^j(\eta^{1\frown}\langle\beta\rangle) = \lambda^j(\lambda(\eta^{1\frown}\langle\beta\rangle)) = \lambda^j(\lambda(\eta^1)) = \lambda^j(\eta^1),$$

so $\eta^{1\frown}\langle\beta\rangle$ is nonswitching.

Suppose that $n > 1$ and that η^1 is the initial derivative of $\text{up}(\eta^1)$ along η^1. By Lemma 8.1.1(i) (Limit Path), $\text{up}(\eta^1) = \lambda(\eta^1)$. If $\text{up}(\eta^1)^\frown\langle\beta^1\rangle$ is nonswitching for every $\beta^1 \in T^1$ such that $\text{up}(\eta^1)^\frown\langle\beta^1\rangle \in T^2$ and $(\beta^1)^-$ has Π outcome (Σ outcome, resp.) along β^1, fix $\beta \in T^0$ such that $\eta^{1\frown}\langle\beta\rangle \in T^1$ and β^- has Σ outcome (Π outcome, resp.) along β; one of these cases must hold by induction, using the relativized lemma for $k = 2$. By Lemma 8.1.1(ii) (Limit Path), $\lambda(\eta^{1\frown}\langle\beta\rangle) = \text{up}(\eta^1)^\frown\langle\eta^{1\frown}\langle\beta\rangle\rangle$. Hence by Lemma 8.1.1(i) (Limit Path),

$$\lambda^j(\eta^{1\frown}\langle\beta\rangle) = \lambda^j(\lambda(\eta^{1\frown}\langle\beta\rangle)) = \\ \lambda^j(\text{up}(\eta^1)^\frown\langle\eta^{1\frown}\langle\beta\rangle\rangle) \supseteq \lambda^j(\text{up}(\eta^1)) = \lambda^j(\lambda(\eta^1)) = \lambda^j(\eta^1)$$

for all $j \in [2, n]$, and so $\eta^{1\frown}\langle\beta\rangle$ is nonswitching.

Suppose that $\eta^{1\frown}\langle\beta\rangle$ is a nonswitching extension of η^1. We note that for all $j \in [1, n]$, $\lambda^j(\eta^{1\frown}\langle\beta\rangle) \supseteq \lambda^j(\eta^1)$. Hence every primary $\lambda^j(\eta^1)$-link is a primary $\lambda^j(\eta^{1\frown}\langle\beta\rangle)$-link, and so every $\lambda^j(\eta^1)$-link is a $\lambda^j(\eta^{1\frown}\langle\beta\rangle)$-link. Fix p and s as in the proof of Lemma 8.1.3 (λ-Behavior). As $\eta^{1\frown}\langle\beta\rangle$ is a nonswitching extension of η^1, it follows from the proof of Lemma 8.1.3 (λ-Behavior) that $p = s$. If $j \leq p$, then by Lemma 8.1.3(i) (λ-Behavior), $(\lambda^j(\eta^{1\frown}\langle\beta\rangle))^- = \lambda^j(\eta^1)$ and $\lambda^j(\eta^1)$ is the initial derivative of $\text{up}(\lambda^j(\eta^1))$ along $\lambda^j(\eta^{1\frown}\langle\beta\rangle)$; and if $j > p$, then by Lemma 8.1.3(iii) (λ-Behavior), $\lambda^j(\eta^{1\frown}\langle\beta\rangle) = \lambda^j(\eta^1)$. In either case, any $\lambda^j(\eta^{1\frown}\langle\beta\rangle)$-link is a $\lambda^j(\eta^1)$-link. ⊣

LEMMA 8.2.2 (Switching Extension Lemma). *Fix $k \leq n$ and $\eta^k \in T^k$. Then one of the following conditions holds:*

(i) $\eta^{k\frown}\langle\beta^{k-1}\rangle$ is switching for all $\beta^{k-1} \in T^{k-1}$ such that $\eta^{k\frown}\langle\beta^{k-1}\rangle \in T^k$ and $(\beta^{k-1})^-$ has Σ outcome along β^{k-1}.

(ii) $\eta^k{}^\frown\langle\beta^{k-1}\rangle$ *is switching for all* $\beta^{k-1} \in T^{k-1}$ *such that* $\eta^k{}^\frown\langle\beta^{k-1}\rangle \in T^k$ *and* $(\beta^{k-1})^-$ *has* Π *outcome along* β^{k-1}.

(iii) η^k *is the initial derivative of* $\mathrm{up}^j(\eta^k)$ *along* $\lambda^j(\eta^k)$ *for all* $j \in [k, n]$.

PROOF. We prove the lemma for $k = 1$, noting that the proof relativizes. We proceed by induction on n. (iii) holds if $n = 1$. Suppose that $n > 1$. Let μ^1 be the initial derivative of $\mathrm{up}(\eta^1)$ along $\lambda(\eta^1)$. If $\mu^1 \subset \eta^1$, then by (2.6.2) and (2.6.3), $\mathrm{up}(\eta^1)$ must have Π outcome along $\lambda(\eta^1)$; so if β is a Π outcome for η^1, then it follows from Definition 2.4.1 that $\lambda(\eta^1) \mid \lambda(\eta^1{}^\frown\langle\beta\rangle)$, and (ii) holds. Suppose that $\mu^1 = \eta^1$. By Lemma 8.1.1(i) (Limit Path), $\mathrm{up}(\eta^1) = \lambda(\eta^1)$. If $\mathrm{up}(\eta^1)^\frown\langle\beta\rangle$ is switching for all $\beta \in T^0$ such that β^- has Σ outcome along β, then (ii) holds. If $\mathrm{up}(\eta^1)^\frown\langle\beta\rangle$ is switching for all $\beta \in T^0$ such that β^- has Π outcome along β, then (i) holds. Otherwise, by induction, $\mathrm{up}(\eta^1)$ is the initial derivative of $\mathrm{up}^j(\mathrm{up}(\eta^1)) = \mathrm{up}^j(\eta^1)$ along $\lambda^j(\lambda(\eta^1)) = \lambda^j(\eta^1)$ for all $j \in [1, n]$, so (iii) holds. ⊣

8.3. Links

Condition (2.6.4) imposes restrictions on the path generation process that are reflected when the extension taken is a switching extension. The switching process removes nodes from the current path by creating links that restrain derivatives of those nodes. In order to return these nodes to a newly computed current path, the links must be removed. We first show that links are nested, so that the process of removing links by switching paths must be carried out in an orderly manner, no matter how a path is extended. We will also determine the relationship between the restraint of η^k by a $\lambda^k(\rho^j)$-link, and the location of antiderivatives of η^k relative to the paths through trees computed by ρ^j, thus analyzing the relationship between links and switching nodes.

The process of returning a node to the current path will proceed by taking extensions that switch the outcomes of the last nodes of links. It will be important to characterize the links that must be switched in this way. The characterization will depend on the fact that any two ρ^k-links are either nested or disjoint.

LEMMA 8.3.1 (Nesting Lemma). *Fix* $k \leq n$ *and* $\rho^k \in T^k \cup [T^k]$. *Suppose that, for* $i \leq 1$, $[\mu_i^k, \pi_i^k]$ *is a* ρ^k*-link and that* $\pi_0^k \subset \pi_1^k$. *Then either* $\mu_1^k \subseteq \mu_0^k$ *or* $\pi_0^k \subset \mu_1^k$.

PROOF. We proceed first by induction on $n - k$, then by induction on $|\rho^k|$, and finally by induction on $|\rho^k| - |\pi_0^k|$. As there are no links on T^n, the lemma holds trivially for $k = n$. Assume that the lemma is true for $k + 1$ in place of k. As the proof relativizes, we just present the proof for $k = 0$, and as is our convention, drop the superscript when it is 0.

For each $i \leq 1$, fix the $\lambda(\rho)$-link $[\mu_i^1, \pi_i^1]$ from which $[\mu_i, \pi_i]$ is derived if such a link exists; otherwise, $[\mu_i, \pi_i]$ is a primary ρ-link, and we set $\mu_i^1 = \pi_i^1 = \text{up}(\mu_i) = \text{up}(\pi_i)$. We note that by Definitions 2.4.4 and 2.5.4, for $i \leq 1$, μ_i (π_i, resp.) is the initial (principal, resp.) derivative of μ_i^1 (π_i^1, resp.) along ρ. First assume that π_0^1 and π_1^1 are comparable. As $\pi_0 \subset \pi_1$, it follows from Lemma 8.1.1(ii) (Limit Path) that $\pi_0^1 \subset \pi_1^1$. If $\pi_0^1 \subset \mu_1^1$, then by Lemma 8.1.1(i),(ii) (Limit Path), $\mu_0 \subset \pi_0 \subset \mu_1 \subset \pi_1$ and the lemma holds. Otherwise, $\mu_1^1 \subseteq \pi_0^1 \subset \pi_1^1$. If $\mu_0^1 = \pi_0^1$, then $\mu_1^1 \subseteq \mu_0^1$. And if $\mu_0^1 \neq \pi_0^1$, then by induction, $\mu_1^1 \subseteq \mu_0^1$. In either case, it follows from Lemma 8.1.1(i) (Limit Path) that $\mu_1 \subseteq \mu_0$.

Now assume that $\pi_0^1 \mid \pi_1^1$, and let $\kappa^1 = \pi_0^1 \wedge \pi_1^1$. By (2.6.1), $\lambda(\pi_i) \supseteq \pi_i^1$ for $i \leq 1$, so by Remark 2.4.2(vi) and as $\pi_0 \subset \pi_1$, κ^1 has a principal derivative μ that has Σ outcome along π_0 and so is the initial derivative of κ^1 along π_0, and a principal derivative $\pi \in (\pi_0, \pi_1)$ along π_1 that has Π outcome along π_1. Thus $[\mu, \pi]$ is a primary π_1-link that restrains π_0, and so by induction on $|\rho|$, $\mu \subseteq \mu_0$. As $[\mu, \pi]$ is a primary π_1-link and $\pi_1 \subset \rho$, $[\mu, \pi]$ is also a primary ρ-link. Hence by induction on $|\rho| - |\pi_0|$, either $\pi \subset \mu_1$ or $\mu_1 \subseteq \mu$. As $\mu \subseteq \mu_0$ and $\pi_0 \subset \pi$, it follows that either $\pi_0 \subset \mu_1$ or $\mu_1 \subseteq \mu_0$. ⊣

The next lemma identifies the outcome of a link with the actual outcome of the node ending the link.

LEMMA 8.3.2 (Faithful Outcome Lemma). *Fix $\mu^k \subset \pi^k \subset \rho^k \in T^k \cup [T^k]$ such that $[\mu^k, \pi^k]$ is an ρ^k-link. Then $[\mu^k, \pi^k]$ has Σ outcome iff π^k has Σ outcome along ρ^k.*

PROOF. We proceed by induction on $n - k$. As there are no links on T^n, the lemma holds for $k = n$. Assume that $k < n$. As the proof relativizes, we just present the proof for $k = 0$, and as is our convention, drop the superscript when it is 0.

First assume that $[\mu, \pi]$ has Σ outcome. Then $[\mu, \pi]$ is not a primary ρ-link, so $[\mu, \pi]$ is derived from some $\lambda(\rho)$-link $[\mu^1, \pi^1]$, and $[\mu^1, \pi^1]$ has Π outcome. By induction, π^1 has Π outcome along $\lambda(\rho)$, so by Definition 2.2.2, π must have Σ outcome along ρ.

Assume that $[\mu, \pi]$ has Π outcome. If $[\mu, \pi]$ is a primary ρ-link, then π has Π outcome along ρ. Otherwise, $[\mu, \pi]$ is derived from some $\lambda(\rho)$-link $[\mu^1, \pi^1]$, and $[\mu^1, \pi^1]$ has Σ outcome. By induction, π^1 has Σ outcome β along $\lambda(\rho)$, so by Definition 2.2.2, β^- has Π outcome along $\beta \subseteq \rho$. By Definition 2.5.4, $[\mu, \pi]$ is the unique ρ-link of T derived from $[\mu^1, \pi^1]$, and $\pi = \beta^-$. So π has Π outcome along ρ. ⊣

The next lemma relates the existence of antiderivatives of a node η^k on a path computed by ρ^k to the restraint of η^k by a ρ^k-link. It is used frequently to determine whether or not links restraining a given node are present.

LEMMA 8.3.3 (Link Analysis Lemma). *Fix* $k \le r \le n$ *and* $\eta^k \subseteq \rho^k \in T^k \cup [T^k]$. *Then:*

(*i*) *If* $\mathrm{up}(\eta^k)|\lambda(\rho^k)$ *and* $\kappa^k = \mathrm{up}(\eta^k) \wedge \lambda(\rho^k)$, *then there are* $\sigma^k \subset \tau^k \subseteq \rho^k$ *such that* $\lambda(\tau^k) \subseteq \lambda(\rho^k)$ *and* $[(\sigma^k)^-, (\tau^k)^-]$ *is a primary* ρ^k*-link that properly* ρ^k*-restrains all* $\xi^k \subset \rho^k$ *for which* $\mathrm{up}(\xi^k) \supset \kappa^k$ *and* $\mathrm{up}(\xi^k)|\lambda(\tau^k)$.

(*ii*) *If* $[\widetilde{\mu}^k, \widetilde{\pi}^k]$ *is an* ρ^k*-link that* ρ^k*-restrains the principal derivative* η^k *of* $\mathrm{up}(\eta^k)$ *along* ρ^k, *then either there is a primary* ρ^k*-link* $[\mu^k, \pi^k]$ *that properly* ρ^k*-restrains* η^k *and* $\mathrm{up}(\eta^k) \not\subseteq \lambda(\rho^k)$, *or* $[\widetilde{\mu}^k, \widetilde{\pi}^k]$ *is derived from a* $\lambda(\rho^k)$*-link that* $\lambda(\rho^k)$*-restrains* $\mathrm{up}(\eta^k)$.

(*iii*) *If there is a primary* ρ^k*-link* $[\mu^k, \pi^k]$ *that* ρ^k*-restrains the initial derivative* α^k *of* $\mathrm{up}(\eta^k)$ *along* ρ^k, *then either* $\mathrm{up}(\eta^k) \not\subseteq \lambda(\rho^k)$ *or* $\mu^k = \alpha^k$.

PROOF. As the proof relativizes, we just present the proof for $k = 0$, and as is our convention, drop the superscript when it is 0.

(i): As $\eta \subset \rho$ and $\mathrm{up}(\eta) \mid \lambda(\rho)$, it follows from Lemma 8.1.1(i),(ii) (Limit Path) that κ^1 has an initial derivative μ and a principal derivative π such that $\mu \subset \pi \subset \rho$. Fix $\sigma, \tau \subseteq \rho$ such that $\sigma^- = \mu$ and $\tau^- = \pi$, and note that by 8.1.1(i),(ii) (Limit Path) and (2.6.2), $\lambda(\sigma) = \kappa^1{}^\frown\langle\sigma\rangle \subseteq \mathrm{up}(\eta)$ and $\lambda(\tau) = \kappa^1{}^\frown\langle\tau\rangle \subseteq \lambda(\rho)$. By Lemma 8.1.1(i,ii) (Limit Path), $\mu \subset \eta \subset \pi$. (i) now follows from the observation that the choice of $[\mu, \pi]$ will be identical for any $\xi \subset \rho$ for which $\mathrm{up}(\xi) \supset \kappa$ and $\mathrm{up}(\xi)|\lambda(\tau)$.

(ii): First suppose that $[\widetilde{\mu}, \widetilde{\pi}]$ is a primary ρ-link. As $\widetilde{\pi}$ is the principal derivative of $\mathrm{up}(\widetilde{\pi})$ along ρ and is not ρ-restrained by $[\widetilde{\mu}, \widetilde{\pi}]$, we cannot have $\mathrm{up}(\widetilde{\pi}) = \mathrm{up}(\eta)$. As the link is primary, $\mathrm{up}(\widetilde{\mu}) = \mathrm{up}(\widetilde{\pi})$; hence the restraint of η by this link must be proper.

Otherwise, $[\widetilde{\mu}, \widetilde{\pi}]$ must derived from the $\lambda(\rho)$-link $[\mathrm{up}(\widetilde{\mu}), \mathrm{up}(\widetilde{\pi})]$. We compare the locations of $\mathrm{up}(\eta)$, $\mathrm{up}(\widetilde{\mu})$ and $\mathrm{up}(\widetilde{\pi})$ on T^1. We cannot have $\mathrm{up}(\eta) \subset \mathrm{up}(\widetilde{\mu})$, else by Lemma 8.1.1(ii) (Limit Path), we would have $\eta \subset \widetilde{\mu}$ so $[\widetilde{\mu}, \widetilde{\pi}]$ would not ρ-restrain η. We cannot have $\mathrm{up}(\widetilde{\pi}) \subseteq \mathrm{up}(\eta)$, else by Lemma 8.1.1(ii) (Limit Path) and as $\widetilde{\pi}$ and η are both principal derivatives along ρ, we would have $\widetilde{\pi} \subseteq \eta$ so $[\widetilde{\mu}, \widetilde{\pi}]$ would not ρ-restrain η. If $\mathrm{up}(\widetilde{\mu}) \subseteq \mathrm{up}(\eta) \subset \mathrm{up}(\widetilde{\pi})$, then $[\mathrm{up}(\widetilde{\mu}), \mathrm{up}(\widetilde{\pi})]$ $\lambda(\rho)$-restrains $\mathrm{up}(\eta)$ and (ii) follows. Otherwise, we must have $\mathrm{up}(\eta)|\,\mathrm{up}(\widetilde{\pi}) \subseteq \lambda(\pi)$, so by Remark 2.4.2(ii), $\mathrm{up}(\eta)|\lambda(\rho)$; (ii) now follows from (i) in this case.

(iii): Suppose that $[\mu, \pi]$ is a primary ρ-link that ρ-restrains the initial derivative α of $\mathrm{up}(\eta)$ along ρ, and that $\mathrm{up}(\eta) \subseteq \lambda(\rho)$. We assume that $\mu \neq \alpha$ and derive a contradiction. By Lemma 8.3.1 (Nesting), there can be no primary α-link that α-restrains μ, so by (i), $\mathrm{up}(\mu) \subseteq \lambda(\alpha)$. By (2.6.1), $\mathrm{up}(\alpha) \subseteq \lambda(\alpha)$, so $\mathrm{up}(\mu)$ and $\mathrm{up}(\alpha)$ are comparable. As $\mu \neq \alpha$, it follows from Lemma 8.1.1(i) (Limit Path) that $\mathrm{up}(\mu) \subset \mathrm{up}(\alpha)$. As $[\mu, \pi]$ is a primary ρ-link and $\alpha \subset \pi$, it follows from (2.6.2) that all derivatives of $\mathrm{up}(\mu)$ that are $\subset \alpha$ must have Σ outcome along α, and π must have Π outcome along ρ, so as $\mu \neq \alpha$, $\mathrm{up}(\mu)$ must have Π outcome along $\lambda(\alpha)$, but Σ outcome along

$\lambda(\rho)$. But $\text{up}(\mu) \subset \text{up}(\alpha) = \text{up}(\eta) \subseteq \lambda(\rho)$, so by (2.6.1), $\text{up}(\alpha) \subseteq \lambda(\alpha), \lambda(\rho)$. Thus $\text{up}(\mu)$ must have the same outcome along both $\lambda(\alpha)$ and $\lambda(\rho)$, yielding a contradiction. $\dashv$

The next two definitions trace links to higher trees, and allow us to identify the levels where the paths computed by η^k are compatible with those computed by ρ^k. A node $\eta^k \subseteq \rho^k$ will be ρ^k-*true* if the paths computed by η^k are compatible with those computed by ρ^k. Recall that $\eta^k \subseteq \rho^k$ is ρ^k-*free* if η^k is not restrained by any link along ρ^k. While these notions are closely related, they are not identical. We will show, in Lemma 8.3.7 (Free and True) that η^k is ρ^k-true iff $(\eta^k)^-$ is ρ^k-free.

DEFINITION 8.3.4. Fix $k \leq r \leq n$ and $\eta^k \subseteq \rho^k \in T^k \cup [T^k]$. We say that η^k is (ρ^k, r)-*true* if $\lambda^j(\eta^k) \subseteq \lambda^j(\rho^k)$ for all $j \in [k, r]$. η^k is ρ^k-*true* if η^k is (ρ^k, n)-true.

DEFINITION 8.3.5. Fix $k \leq r \leq n$ and $\eta^k \subset \rho^k \in T^k \cup [T^k]$. Let $[\mu^k, \pi^k]$ be a ρ^k-link. We say that $[\mu^k, \pi^k]$ is a (ρ^k, r)-*link* if $[\mu^k, \pi^k]$ is derived from a primary $\lambda^r(\rho^k)$-link. $[\mu^k, \pi^k]$ is a $(\rho^k, \leq r)$-*link* if $[\mu^k, \pi^k]$ is a (ρ^k, j)-link for some $j \in [k, r]$. We say that η^k is $(\rho^k, \leq r)$-*restrained* if η^k is restrained by a $(\rho^k, \leq r)$-link, and that η^k is (ρ^k, r)-*restrained* if η^k is restrained by a (ρ^k, r)-link. η^k is (ρ^k, r)-*free* if η^k is not (ρ^k, r)-restrained.

We now show that ρ^k-links are preserved by $\tau^k \supseteq \rho^k$, as long as ρ^k is τ^k-true.

LEMMA 8.3.6 (Link Preservation Lemma). *Fix $k \leq r \leq n$, $\tau^k \in T^k \cup [T^k]$, and $\mu^k \subset \pi^k \subset \rho^k$ such that ρ^k is (τ^k, r)-true. Then $[\mu^k, \pi^k]$ is a (τ^k, r)-link iff $[\mu^k, \pi^k]$ is an (ρ^k, r)-link.*

PROOF. As the proof relativizes, we just present the proof for $k = 0$, and as is our convention, drop the superscript when it is 0.

Clearly, any (ρ, r)-link is a (τ, r)-link. Suppose that $[\mu, \pi]$ is a (τ, r)-link and that $\pi \subset \rho$. For all $j \leq r$, let $[\mu^j, \pi^j]$ be the $\lambda^j(\tau)$-link from which $[\mu, \pi]$ is derived. We proceed by induction on $j \leq r$ showing that $\pi^j \subset \lambda^j(\rho)$; the lemma then follows by definition.

The induction hypothesis holds if $j = 0$. Suppose that $j > 0$. As $j \leq r$ and ρ is (τ, r)-true, we must have $\lambda^j(\rho) \subseteq \lambda^j(\tau)$. As $[\mu^j, \pi^j]$ is a $\lambda^j(\tau)$-link, $\pi^j \subset \lambda^j(\tau)$. Hence π^j and $\lambda^j(\rho)$ are comparable. Fix the immediate successor ν^{j-1} of π^{j-1} along $\lambda^{j-1}(\tau)$. Now by induction and hypothesis, $\pi^{j-1} \subset \lambda^{j-1}(\rho) \subseteq \lambda^{j-1}(\tau)$, and by Definition 2.5.4, π^{j-1} is the principal derivative of π^j along $\lambda^j(\tau)$. Hence by Lemma 8.1.1(ii) (Limit Path), $\pi^{j\frown}\langle \nu^{j-1}\rangle \subseteq \lambda^j(\rho)$. $\dashv$

For a given level n construction with true path $\Lambda^0 \in [T^0]$, the nodes $\eta \subset \Lambda^0$ that ensure that all requirements assigned to nodes $\eta^n \subset \Lambda^n = \lambda^n(\Lambda^0)$ are satisfied are the Λ^0-free nodes. In order to show that this is the case, we will need to know that the finite strings determined by action along the immediate successors ν along Λ^0 of such η are initial segments on the final sets to which they are approximating along Λ^0. This will normally follow if we can show that $\lambda^j(\nu) \subset \lambda^j(\Lambda^0)$ for all $j \leq n$; in other words, we will need to show that if η is Λ^0-free, then ν is Λ^0-true. The converse of this statement will also hold. These conditions will follow from our next lemma.

LEMMA 8.3.7 (Free and True Lemma). *Fix $k < r \leq n$ and $\nu^k \subseteq \rho^k \in T^k \cup [T^k]$. Then:*

(i) *If ν^k is not (ρ^k, r)-true, then $(\nu^k)^-$ is $(\rho^k, \leq r-1)$-restrained.*
(ii) *ν^k is ρ^k-true iff $(\nu^k)^-$ is ρ^k-free.*
(iii) *If ν^k is ρ^k-free, then* $\text{up}(\nu^k)$ *is $\lambda(\rho^k)$-free. In fact, if* $\text{up}(\nu^k)$ *is $\lambda(\rho^k)$-restrained by a $\lambda(\rho^k)$-link $[\mu^{k+1}, \pi^{k+1}]$ derived from a primary $\lambda^r(\rho^k)$-link, then either ν^k is ρ^k-restrained by the ρ^k-link derived from $[\mu^{k+1}, \pi^{k+1}]$, or ν^k is ρ^k-restrained by a ρ^k-link derived from a primary $\lambda^s(\rho^k)$-link for some $s < r$.*

PROOF. As the proof relativizes, we just present the proof for $k = 0$, and as is our convention, drop the superscript when it is 0.

(i): Fix the smallest r such that ν is not (ρ, r)-true; it suffices to show that ν^- is $(\rho, \leq r-1)$-restrained. By Lemma 8.3.3(i) (Link Analysis), there is a primary $\lambda^{r-1}(\rho)$-link $[\mu^{r-1}, \pi^{r-1}]$ such that $\mu^{r-1} \subset \lambda^{r-1}(\nu) \subseteq \pi^{r-1}$ and if ξ^{r-1} is the immediate successor of π^{r-1} along $\lambda^{r-1}(\rho)$, then $\xi^{r-1} \subseteq \lambda^{r-1}(\rho)$. For all $j \leq r-1$ let $[\mu^j, \pi^j]$ be the $\lambda^j(\rho)$-link derived from $[\mu^{r-1}, \pi^{r-1}]$, and let ξ^j be the immediate successor of π^j along $\lambda^j(\rho)$. By Lemma 8.1.1(i,ii) (Limit Path), for all $i, j \leq r-1$ such that $i < j$, $\lambda^j(\mu^i) = \mu^j$ and $\lambda^j(\xi^i) = \xi^j$, and by Remark 2.4.2(v), $\mu^i \subset \lambda^i(\nu) \subset \xi^i$. Thus $\mu \subseteq \nu^- \subset \pi$, so ν^- is $(\rho, \leq r-1)$-restrained.

(ii): We prove the contrapositives. For all $i \leq n$, let $\nu^i = \lambda^i(\nu)$ and $\rho^i = \lambda^i(\rho)$. By (i), if ν is not ρ-true, then ν^- is not ρ-free.

Suppose that $\eta = \nu^-$ is not ρ-free. We may assume that $\nu^i \subseteq \rho^i$ for all $i \leq n$, else (ii) is immediate. Fix the smallest j such that η is (ρ, j)-restrained, and fix a primary ρ^j-link $[\mu^j, \pi^j]$ such that η is restrained by a ρ-link $[\mu, \pi]$ derived from $[\mu^j, \pi^j]$; note that $j < n$ as there are no links on T^n. Then $\mu \subseteq \eta \subset \pi$, so $\mu \subset \nu \subseteq \pi$. For all $i \leq j$, let $[\mu^i, \pi^i]$ be the ρ^i-link derived from $[\mu^j, \pi^j]$. As $\nu^i, \pi^i \subseteq \rho^i$ and $\mu^i \subset \pi^i$ for all $i \leq j$, ν^i, π^i and μ^i are all comparable. Hence by Remark 2.4.2(v), Lemma 8.1.1(ii) (Limit Path), and induction on j, $\mu^i \subset \nu^i \subseteq \pi^i$ for all $i \leq j$.

As $[\mu^j, \pi^j]$ is a primary ρ^j-link, it must be the case that μ^j and π^j are, respectively, the initial and principal derivatives of $\text{up}(\mu^j) = \text{up}(\pi^j)$ along ρ^j. By Remark 2.4.2(ii,vi) and (2.6.4), $\text{up}(\mu^j)$ must have Π outcome along ν^{j+1}; and as $\pi^j \subset \rho^j$ and π^j is the principal, but not the initial, derivative of $\text{up}(\mu^j)$ along ρ^j, if $\text{up}(\mu^j) \subseteq \rho^{j+1}$, then $\text{up}(\mu^j)$ has Σ outcome along ρ^{j+1}. Thus $\nu^{j+1} \not\subseteq \rho^{j+1}$, and so ν is not ρ-true.

(iii): For all $j \leq r$, let $\kappa^j = \lambda^j(\nu) \wedge \lambda^j(\rho)$. We may assume that the ρ-link $[\mu, \pi]$ derived from $[\mu^1, \pi^1]$ does not ρ-restrain ν, else we are done. Thus we must have $\pi \subseteq \nu$, and since $[\mu^1, \pi^1]$ $\lambda(\rho)$-restrains $\nu^1 = \mathrm{up}(\nu)$ and hence $\nu^1 \subset \pi^1$, we conclude that $\pi \subset \nu$.

Fix the smallest $p \leq r$, if any, such that $\lambda^p(\nu)|\lambda^p(\rho)$; such a p must exist, else $[\mu^1, \pi^1]$ would also be a $\lambda(\nu)$-link, so ν^1 would not be $\lambda(\nu)$-free contrary to (2.6.4). By Remark 2.4.2(iv), κ^p must have an initial derivative $\widetilde{\mu}^{p-1} \subset \lambda^{p-1}(\nu)$ with Σ outcome along $\lambda^{p-1}(\nu)$, and a principal derivative $\widetilde{\pi}^{p-1} \in [\lambda^{p-1}(\nu), \lambda^{p-1}(\rho))$ with Π outcome along $\lambda^{p-1}(\rho)$, and so $[\widetilde{\mu}^{p-1}, \widetilde{\pi}^{p-1}]$ is a primary $\lambda^{p-1}(\rho)$-link. For all $j < p$, let $[\widetilde{\mu}^j, \widetilde{\pi}^j]$ be the $\lambda^j(\rho)$-link derived from $[\widetilde{\mu}^{p-1}, \widetilde{\pi}^{p-1}]$, and fix $\xi^j \subseteq \lambda^j(\rho)$ such that $(\xi^j)^- = \widetilde{\pi}^j$. (As is our convention, we omit the superscript 0.) (iii) will follow once we show that $[\widetilde{\mu}, \widetilde{\pi}]$ ρ-restrains ν.

By Lemma 8.1.1(i) (Limit Path), $\widetilde{\mu}^j \subset \lambda^j(\nu)$ for all $j \leq p-1$. Furthermore, by choice of p and as $\widetilde{\pi}^j \subset \lambda^j(\rho)$ for all $j \leq p$, $\widetilde{\pi}^j$, ξ^j and $\lambda^j(\nu)$ must be comparable for all $j \leq p-1$. We compare the locations of $\lambda^j(\nu)$, ξ^j and $\widetilde{\pi}^j$ for $j < p$.

First assume that there is a $j < p$ such that $\lambda^j(\nu) \supseteq \xi^j$. If $j < p-1$, then by Lemma 8.1.1(i) (Limit Path) and Definition 2.4.1 and since $\lambda^{j+1}(\nu)$ and ξ^{j+1} are comparable and $\lambda^{j+1}(\nu) \subseteq \lambda^{j+1}(\rho)$, we must have $\lambda^{j+1}(\nu) \supseteq \xi^{j+1}$; hence by induction, $\lambda^{p-1}(\nu) \supseteq \xi^{p-1}$. Thus $[\widetilde{\mu}^{p-1}, \widetilde{\pi}^{p-1}]$ is a primary $\lambda^{p-1}(\nu)$-link. But this implies that $\widetilde{\pi}^{p-1} \subset \lambda^{p-1}(\nu)$, a contradiction.

Next assume that there is a $j < p$ such that $\lambda^j(\nu) \subset \widetilde{\pi}^j$. If $j > 0$, then by Lemma 8.1.1(i) (Limit Path) and Definition 2.4.1 and since $\lambda^{j+1}(\nu)$ and ξ^{j+1} are comparable and $\lambda^{j+1}(\nu) \subseteq \lambda^{j+1}(\rho)$, we must have $\lambda^{j-1}(\nu) \subset \widetilde{\pi}^{j-1}$. Hence by induction, $\nu = \lambda^0(\nu) \subset \widetilde{\pi}$, and so $[\widetilde{\mu}, \widetilde{\pi}]$ is a ρ-link restraining ν. As $p - 1 < r$, (iii) now follows in this case.

Otherwise, we must have $\lambda^j(\nu) = \widetilde{\pi}^j$ for all $j \leq p - 1$, so as $\widetilde{\pi}^j$ is a derivative of $\widetilde{\pi}^{j+1}$ for all $j < p - 1$, it follows from Definition 2.4.1 that $\widetilde{\pi}^j$ is the initial derivative of $\widetilde{\pi}^{j+1}$ for all $j < p - 1$. As $\nu = \lambda^0(\nu) = \widetilde{\pi}$, we must have $\nu^j = \mathrm{up}^j(\nu) = \widetilde{\pi}^j$ for all $j \leq p - 1$. Now $[\mu^1, \pi^1]$ and $[\widetilde{\mu}^1, \widetilde{\pi}^1]$ are both $\lambda^1(\rho)$-links, and as $\pi \subset \nu = \widetilde{\pi}$ and both π and $\widetilde{\pi}$ are principal derivatives, it follows from Lemma 8.1.1(ii) (Limit Path) that $\pi^1 \subset \widetilde{\pi}^1$. But then $\nu^1 \subset \pi^1 \subset \widetilde{\pi}^1 = \nu^1$, a contradiction. Thus this case cannot occur, and the lemma follows. ⊣

8.4. Free Nodes

We now prove two lemmas that specify that certain nodes are free. The first lemma will be used to analyze the path-switching process; it will be used to extend paths, thereby returning a node that is not along the current path to the current path as computed by the extension. The second lemma will allow us to conclude that free nodes have sufficiently many free derivatives. Our inductive process requires nodes to pass down the responsibility for satisfying a given condition to its free derivatives, each of which will have the responsibility to satisfy instances of that condition. This lemma will be used to show that if the derivatives succeed in satisfying their conditions, then the condition for the original node will be satisfied.

When a requirement is assigned to $\eta^k \in T^k$, then by Lemma 8.3.7(iii) (Free and True) and the process of pulling links down from tree to tree, $\text{up}^i(\eta^k)$ is $\lambda^i(\eta^k)$-free for all $i \in [k, n]$. If $(\nu^k)^- = \eta^k$ and $i \in [k, n]$, i.e., ν^k determines an outcome for η^k, then ν^k may or may not switch $\text{up}^i(\eta^k)$. We show that in either case, no $\lambda^i(\nu^k)$-link restrains $\text{up}^i(\eta^k)$ for any $i \in [k, n]$. Furthermore, if ν^k is r-switching, then ν^k switches $\text{up}^i(\eta^k)$ for all $i \in [r, n]$.

LEMMA 8.4.1 (Free Extension Lemma). *Fix $k \le n$ and $\eta^k, \nu^k \in T^k$ such that $\eta^k = (\nu^k)^-$. For all $i \in [k, n]$, let $\eta^i = \text{up}^i(\eta^k)$. Then:*

(i) *For all $i \in [k, n]$, η^i is $\lambda^i(\nu^k)$-free.*

(ii) *If ν^k is r-switching, then ν^k switches η^i for all $i \in [r, n]$, and each such switch interchanges Σ outcomes and Π outcomes.*

PROOF. As the proof relativizes, we just present the proof for $k = 0$, and as is our convention, drop the superscript when it is 0. Fix p and s as in Lemma 8.1.3 (λ-Behavior).

(i): If $i \le s$, then by Lemma 8.1.3(i,ii) (λ-Behavior), $\eta^i = (\lambda^i(\nu))^-$. If $[\mu^i, \pi^i]$ is a $\lambda^i(\nu)$-link, then π^i is a principal derivative along $\lambda^i(\nu)$, so $\pi^i \subseteq (\lambda^i(\nu))^- = \eta^i$. But $[\mu^i, \pi^i]$ does not restrain any node $\supseteq \pi^i$, so (i) follows in this case.

Suppose that $i > s$. By Lemma 8.1.3(ii,iii) (λ-Behavior), $\lambda^i(\nu) = \lambda^i(\eta^s)$, and by (2.6.4) and Lemma 8.3.7(iii) (Free and True), η^i is $\lambda^i(\eta^s)$-free, so (i) again follows.

(ii): Suppose that ν is r-switching. By Lemma 8.1.3 (λ-Behavior), $r = p+1$. We proceed by induction on $i \in [p+1, n]$. By Lemma 8.1.3(ii) (λ-Behavior) and Remark 2.4.2(vi), ν switches η^{p+1} from Π outcome to Σ outcome.

Suppose that $i > p+1$. If η^{i-1} has Π outcome along $\lambda^{i-1}(\nu)$, then by (i) and Definition 2.4.1, η^i has Σ outcome along $\lambda^i(\nu)$. By induction, ν switches η^{i-1} and η^{i-1} has Σ outcome along $\lambda^{i-1}(\eta)$. By (2.6.4) and Lemma 8.3.7(iii) (Free and True), η^{i-1} is $\lambda^{i-1}(\eta)$-free, so by (2.6.2), each derivative of η^i along $\lambda^{i-1}(\eta)$ must have Σ outcome along $\lambda^{i-1}(\eta)$. Hence by Remark 2.4.2(vi), η^i has Π outcome along $\lambda^i(\eta)$.

On the other hand, suppose that η^{i-1} has Σ outcome along $\lambda^{i-1}(\nu)$. By (i), η^{i-1} is $\lambda^{i-1}(\nu)$-free. Hence by (2.6.2), all derivatives of η^i along $\lambda^{i-1}(\nu)$ must have Σ outcome along $\lambda^{i-1}(\nu)$, so by Remark 2.4.2(vi), η^i has Π outcome along $\lambda^i(\nu)$. Now by induction, η^{i-1} has Π outcome along $\lambda^{i-1}(\eta)$, and by (2.6.4) and Lemma 8.3.7(iii) (Free and True), $\eta^i \subset \lambda^i(\eta)$; hence by Remark 2.4.2(vi), η^i has Σ outcome along $\lambda^i(\eta)$. ⊣

The nodes that are Λ^k-free have the responsibility to determine activated and validated action. However, we will be unable to effectively identify these nodes, and so, will be unable to prevent other nodes from acting. We will have to show that the nodes to which we want to assign responsibility for action can automatically transfer this responsibility to their derivatives. In order for this transfer to occur, we will need to show that principal derivatives of free nodes are free, and that if a free node has Π outcome along the true path Λ^{k+1} through T^{k+1}, then it has infinitely many free derivatives along the true path Λ^k through T^k. We show this in our next lemma.

LEMMA 8.4.2 (Free Derivative Lemma). *Fix $k < n$ and $\rho^k \in T^k \cup [T^k]$. For all $j \in [k, n]$, let $\rho^j = \lambda^j(\rho^k)$. Suppose that $\eta^{k+1} \subset \rho^{k+1}$ is ρ^{k+1}-free. Then:*

(*i*) *The principal derivative $\eta^k \subset \rho^k$ of η^{k+1} along ρ^k is ρ^k-free.*

(*ii*) *If η^{k+1} has Π outcome along ρ^{k+1}, $\text{lev}(\eta^{k+1}) > k$ and $|\rho^k| = \infty$, then η^{k+1} has infinitely many ρ^k-free derivatives.*

PROOF. As the proof relativizes, we just present the proof for $k = 0$, and as is our convention, drop the superscript when it is 0.

(i): By Lemma 8.1.1(ii) (Limit Path), fix the principal derivative η of η^1 along ρ. Suppose that η is ρ-restrained by a ρ-link $[\mu, \pi]$ in order to obtain a contradiction. Without loss of generality, we may assume that $[\mu, \pi]$ is a primary ρ-link, else by Lemma 8.3.7(iii) (Free and True), η^1 would be ρ^1-restrained by a ρ^1-link, so would not be ρ^1-free, contrary to hypothesis. By Lemma 8.3.1(Nesting), we may assume that π is not ρ-restrained. Let ν be the immediate successor of η along ρ. By Lemma 8.1.1(ii) (Limit Path), $(\lambda(\nu))^- = \text{up}(\eta)$ and by Lemma 8.3.3(i) (Link Analysis) and Lemma 8.1.1(ii) (Limit Path), $\lambda(\nu) \subseteq \lambda(\rho)$. By Lemma 8.3.7(ii) (Free and True), $\lambda(\nu)$ is $\lambda(\rho)$-true. Hence ν is ρ-true, so by Lemma 8.3.7(ii) (Free and True), η is ρ-free.

(ii): We note by Lemma 8.1.1(i) (Limit Path) that if $\zeta^n \subset \rho^n$, then for all $j \leq n$, the initial derivative ζ^j of ζ^n along ρ^j is ρ^j-true. As $|\rho| = \infty$, it follows from Definition 2.7.5 that ζ completes a ρ-path through a block. If suffices to show that for all such ζ for which $\text{up}^n(\eta) \subset \zeta^n$, η^1 has a ρ-free derivative in the block, B, containing ζ. Fix such a ζ.

By Lemma 8.3.7(iii) (Free and True), for all $j \in [1, n]$, η^j is ρ^j-free, so by Lemma 8.3.6 (Link Preservation), η^j is ζ^j-free. By Lemma 8.1.1(i) (Limit Path), $\lambda(\zeta^j) = \zeta^{j+1}$ for all $j < n$. It now follows easily from our hypothesis that η^1 is ζ-consistent. By clause (ii) of Definition 2.7.7, η^1 has a ζ-free derivative ξ in B. As ζ is ρ-true, it follows from Lemma 8.3.6 (Link Preservation) that ξ is ρ-free. ⊣

8.5. Relationships Between Links

Constructions starting on trees of arbitrarily high finite levels will require a more detailed analysis of the framework. Many of the facts that will be needed deal with an analysis of how links are created and destroyed. Results of that type are presented in this section; their ultimate value will be in their use in subsequent chapters. We fix a number ℓ as the level of a given construction, and refer to that number throughout this section. Our first lemma summarizes the facts we will need about the freeness of nodes.

LEMMA 8.5.1 (Freeness Transmission Lemma). *Fix $k \leq \ell, \kappa^k, \sigma^k \in T^k$ and $\eta \in T^0$ such that σ^k is $\lambda^k(\eta)$-free and $\kappa^k = \lambda^k(\eta^-) \wedge \lambda^k(\eta)$. Then*:

(*i*) *For all $j \in (k, \ell]$, $\text{up}^j(\sigma^k)$ is $\lambda^j(\eta)$-free.*

(*ii*) *For all $j \leq k$, the principal derivative of σ^k along $\lambda^j(\eta)$ is $\lambda^j(\eta)$-free.*

(*iii*) *κ^k is both $\lambda^k(\eta^-)$-free and $\lambda^j(\eta)$-free.*

PROOF. (i) follows inductively from Lemma 8.3.7(iii) (Free and True). (ii) follows immediately from Lemma 8.4.2(i) (Free Derivative).

Lemma 8.4.1(i) (Free Extension) and Lemma 8.3.1(i) (λ-Behavior) imply that $\kappa^k = \text{up}^k(\eta^-)$ is $\lambda^k(\eta)$-free. As no η^--link can restrain η^-, η^- is η^--free. It now follows from (i) that κ^k is $\lambda^k(\eta^-)$-free. ⊣

The next lemma is technical in nature and will be used to compare links on the same tree. Its clauses will be applied either to the beginnings or ends of links. The main facts to be extracted from this lemma are that if a node properly extends the end of a link on T^k, then its derivatives on T^{k-1} must properly extend the end of the derived link; and when links have non-empty intersection and end at different nodes, then we can specify which of the links is nested within the other.

LEMMA 8.5.2 (Link Comparison Lemma). *Fix $k \leq \ell$ and $\eta_0^k \subseteq \eta_1^k \subset \eta_2^k \subset \nu^k \in T^k$ such that $(\nu^k)^- = \eta_2^k$, For all $j \leq k$ and $i \leq 2$, let $\nu^j = \text{out}^j(\nu^k)$, $\mu_i^j = \text{init}^j(\eta_i^k, \nu^j)$, and $\pi_i^j = \text{prin}^j(\eta_i^k, \nu^j)$. Then for all $j \leq k$,*

$$\mu_0^j \subseteq \mu_1^j \subseteq \pi_1^j \subset \mu_2^j \subseteq \pi_2^j = (\nu^j)^-.$$

PROOF. By Lemma 8.1.1(i) (Limit Path) and the definition of initial derivatives, $\mu_0^j \subseteq \mu_1^j \subseteq \pi_1^j$ and $\mu_2^j \subseteq \pi_2^j$. By Lemma 8.1.1(ii) (Limit Path), $\pi_2^j = (\nu^j)^-$.

It remains to show that $\pi_1^j \subset \mu_2^j$ for all $j \leq k$. As $\eta_1^k \subset \eta_2^k$, we can fix $\nu_1^k \subseteq \eta_2^k$ such that $(\nu_1^k)^- = \eta_1^k$. By Lemma 8.1.1(i) (Limit Path), $\lambda^k(\mu_2^j) = \mu_2^k$ for all $j \leq k$, and by Remark 2.4.2(i), $\nu_1^j = \text{out}^j(\nu_1^k) \subseteq \mu_2^j$ for all $j \leq k$. Furthermore, by Lemma 8.1.2(i) (Out), $(\nu_1^j)^- = \pi_1^j$ for all $j \leq k$. Thus we have shown that $\pi_1^j \subset \mu_2^j$ for all $j \leq k$. ⊣

The next lemma describes the capture of derivatives $\sigma^{k-1} \subset \lambda^{k-1}(\eta)$ of a node $\sigma^k \not\subset \lambda^k(\eta)$ by a primary link. It describes a situation relating two comparable nodes of T^0 that need not be consecutive.

LEMMA 8.5.3 (Crossover Lemma). *Fix $k \leq \ell$, $\nu, \eta \in T^0$, and for all $j \leq \ell$, let $\kappa^j = \lambda^j(\nu) \wedge \lambda^j(\eta)$. Suppose that $\kappa^k \subset \delta^k \subseteq \lambda^k(\nu)$ and that there is* *a* $\delta^{k-1} \subset \kappa^{k-1}$ *such that* $\text{up}(\delta^{k-1}) = \delta^k$. *Then* κ^k *has an initial derivative* $\mu^{k-1} \subset \kappa^{k-1}$ *and a principal derivative* $\pi^{k-1} \subset \lambda^{k-1}(\eta)$ *such that* $[\mu^{k-1}, \pi^{k-1}]$ *is a primary* $\lambda^{k-1}(\eta)$*-link that properly restrains all* $\sigma^{k-1} \subseteq \lambda^{k-1}(\eta)$ *such that* $\kappa^k \subset \text{up}(\sigma^{k-1}) \subset \lambda^k(\nu)$.

PROOF. By Lemma 8.1.1(i) (Limit Path), κ^k has an initial derivative $\mu^{k-1} \subset \delta^{k-1} \subset \kappa^{k-1} \subset \lambda^{k-1}(\nu)$. Let ρ^{k-1} be the immediate successor of μ^{k-1} along κ^{k-1}. As $\kappa^{k-1} \subset \lambda^{k-1}(\eta)$, μ^{k-1} must also be the initial derivative of κ^k along $\lambda^{k-1}(\eta)$, and $\rho^{k-1} \subseteq \lambda^{k-1}(\eta)$.

Let τ^{k-1} be the principal derivative of κ^k along $\lambda^{k-1}(\nu)$. By Lemma 8.5.2 (Link Comparison), $\tau^{k-1} \subset \delta^{k-1} \subset \kappa^{k-1}$. Fix $\gamma^{k-1} \subseteq \delta^{k-1}$ such that $(\gamma^{k-1})^- = \tau^{k-1}$. Now τ^{k-1} cannot have Π outcome along δ^{k-1}, else by (2.6.2) and Remark 2.4.2(i), $\kappa^{k\frown}\langle\gamma^{k-1}\rangle \subseteq \lambda^k(\nu), \lambda^k(\eta)$, contrary to the definition of κ^k. We conclude that $\tau^{k-1} = \mu^{k-1}$ which has Σ outcome along δ^{k-1}, so $\kappa^{k\frown}\langle\rho^{k-1}\rangle \subseteq \lambda^k(\nu)$.

As $\lambda^k(\nu) \wedge \lambda^k(\eta) = \kappa^k$ and $\kappa^{k\frown}\langle\rho^{k-1}\rangle \not\subseteq \lambda^k(\eta)$, it follows from Remark 2.4.2(i) that there must be a $\pi^{k-1} \subset \lambda^{k-1}(\eta)$ such that $\text{up}(\pi^{k-1}) = \kappa^k$ and π^{k-1} has Π outcome along $\lambda^{k-1}(\eta)$. As μ^{k-1} has Σ outcome along $\lambda^{k-1}(\eta)$ and π^{k-1} has Π outcome along $\lambda^{k-1}(\eta)$, it follows from (2.6.2) that $\mu^{k-1} \subset \pi^{k-1}$, so $[\mu^{k-1}, \pi^{k-1}]$ is a primary $\lambda^{k-1}(\eta)$-link. Now if $\pi^{k-1} \subset \zeta^{k-1} \subset \lambda^{k-1}(\eta)$, then by Lemma 8.1.1(ii) (Limit Path), δ^k is not $\lambda(\zeta^{k-1})$-true, so by (2.6.1), δ^k has no derivatives in the interval $(\pi^{k-1}, \lambda^{k-1}(\eta))$. As $\text{up}(\mu^{k-1}) = \text{up}(\pi^{k-1}) = \kappa^k \subset \delta^k$, it follows from the preceding sentence and Lemma 8.1.1(i) (Limit Path) that all derivatives of δ^k along $\lambda^{k-1}(\eta)$ must lie in the interval (μ^{k-1}, π^{k-1}) so are properly $\lambda^{k-1}(\eta)$-restrained by the primary $\lambda^{k-1}(\eta)$-link $[\mu^{k-1}, \pi^{k-1}]$. This link is independent of the choice of δ^k, so the lemma follows. ⊣

As we pass from η^- to η on T^0, new links may be formed, thereby destroying the $\lambda^k(\eta)$-freeness of nodes $\tau^k \in T^k$ that were $\lambda^k(\eta^-)$-free. The manner in which this happens is captured by the next definition, and delineated by the ensuing lemma.

DEFINITION 8.5.4. Fix $i, k, r \leq \ell, \eta \in T^0$ such that $|\eta| > 0$, a primary $\lambda^r(\eta)$-link $[\mu^r, \pi^r]$ and $\tau^k \in T^k$ such that $\tau^k \subseteq \lambda^k(\eta^-)$. We say that $[\mu^r, \pi^r]$ *i-envelops* τ^k *at* η if i is the largest integer $\leq r$ such that the $\lambda^i(\eta)$-link $[\mu^i, \pi^i]$ derived from $[\mu^r, \pi^r]$ $\lambda^i(\eta)$-restrains $\text{up}^i(\tau^k)$ if $i \geq k$ and restrains $\text{prin}^i(\tau^k, \lambda^i(\eta^-))$ if $i < k$.

LEMMA 8.5.5 (Link Development Lemma). *Fix* $i, k, r \leq \ell$, $\mu^r, \pi^r \in T^r$, $\tau^k \in T^k$ *and* $\eta \in T^0$ *such that* $|\eta| > 0$ *and* τ^k *is* $\lambda^k(\eta^-)$*-free. Let* $\kappa^r = \lambda^r(\eta^-) \wedge \lambda^r(\eta)$. *Then*:

(*i*) *If* $[\mu^r, \pi^r]$ *is a primary* $\lambda^r(\eta)$*-link but is not a* $\lambda^r(\eta^-)$*-link, then either* $\pi^r = \kappa^r$ *or* $\kappa^r \subset \mu^r \subset \pi^r \subset \lambda^r(\eta)$.

(*ii*) *If* $[\mu^r, \pi^r]$ *is a primary* $\lambda^r(\eta^-)$*-link but is not a* $\lambda^r(\eta)$*-link, then either* $\pi^r = \kappa^r$ *or* $\kappa^r \subset \mu^r \subset \pi^r \subset \lambda^r(\eta^-)$.

(iii) *If* $[\mu^r, \pi^r]$ *is a primary* $\lambda^r(\eta)$*-link which i-envelops* τ^k *and* $\pi^r \neq \kappa^r$, *then there is a* $j < r$ *and a primary* $\lambda^j(\eta)$*-link* $[\widetilde{\mu}^j, \widetilde{\pi}^j]$ *such that* $\widetilde{\pi}^j = \kappa^j$ *and* $\widetilde{\mu}^j \subset \mu^j \subset \pi^j \subset \widetilde{\pi}^j$, *where* $[\mu^j, \pi^j]$ *is the* $\lambda^j(\eta)$*-link derived from* $[\mu^r, \pi^r]$.

PROOF. (i): Since $[\mu^r, \pi^r]$ is a primary $\lambda^r(\eta)$-link, we must have $\mu^r \subset \pi^r \subset \lambda^r(\eta)$. By Lemma 8.5.1(iii) (Freeness Transmission), κ^r is $\lambda^r(\eta)$-free. Hence either $\pi^r \subseteq \kappa^r$ or $\kappa^r \subset \mu^r$. But we cannot have $\pi^r \subset \kappa^r$, else $[\mu^r, \pi^r]$ would be a $\lambda^r(\eta^-)$-link.

(ii): As in the proof of (i), interchanging η and η^-.

(iii): As $\pi^r \neq \kappa^r$, it follows from (i), Lemma 8.5.1(i,ii) (Freeness Transmission) and the $\lambda^k(\eta^-)$-freeness of τ^k that $\kappa^r \subset \mu^r$. For all $j \leq r$, let $[\mu^j, \pi^j]$ be the $\lambda^j(\eta)$-link derived from $[\mu^r, \pi^r]$. (iii) now follows from Lemma 8.5.3 (Crossover) and the fact that $\pi^0 \subset \eta^-$. ⊣

The next two lemmas specify conditions under which we can convert a $\lambda^k(\eta)$-link to a $\lambda^k(\xi)$-link.

LEMMA 8.5.6 (Reflection Lemma). *Fix* $k \leq \ell$, $\xi, \eta \in T^0$ *and* $\mu^k \subset \pi^k \in T^k$ *such that* $[\mu^k, \pi^k]$ *is a* $\lambda^k(\eta)$*-link,* π^k *is both* $\lambda^k(\eta)$*-free and* $\lambda^k(\xi)$*-free, and* π^k *has* Σ *outcome along* $\lambda^k(\xi)$ *iff* π^k *has* Σ *outcome along* $\lambda^k(\eta)$. *Then* $[\mu^k, \pi^k]$ *is a* $\lambda^k(\xi)$*-link.*

PROOF. Let $[\mu^k, \pi^k]$ be derived from the primary $\lambda^r(\eta)$-link $[\mu^r, \pi^r]$. We proceed by induction on $r - k$. If $r = k$, then $[\mu^r, \pi^r]$ is a primary $\lambda^r(\eta)$-link, so π^r has Π outcome along $\lambda^r(\eta)$. By hypothesis, π^r has Π outcome along $\lambda^r(\xi)$, so $[\mu^r, \pi^r]$ is a primary $\lambda^r(\xi)$-link.

Suppose that $k < r$. As π^k is both $\lambda^k(\eta)$-free and $\lambda^k(\xi)$-free, it follows from Lemma 8.5.1(i) (Freeness Transmission) that up(π^k) must be both $\lambda^{k+1}(\xi)$-free and $\lambda^{k+1}(\eta)$-free; and it follows from Remark 2.4.2(i) that if π^k has Σ (Π, resp.) outcome along both $\lambda^k(\eta)$ and $\lambda^k(\xi)$ then up(π^k) has Π (Σ, resp.) outcome along both $\lambda^{k+1}(\xi)$ and $\lambda^{k+1}(\eta)$. Furthermore, [up(μ^k), up(π^k)] is a $\lambda^{k+1}(\eta)$-link. By induction, [up(μ^k), up(π^k)] is a $\lambda^{k+1}(\xi)$-link, so its derived link $[\mu^k, \pi^k]$ is a $\lambda^k(\xi)$-link. ⊣

LEMMA 8.5.7 (Stable Link Lemma). *Fix* $k \leq \ell$, $\xi, \eta \in T^0$, *and* $\mu^k \subset \pi^k \subseteq \kappa^k \in T^k$ *such that* $[\mu^k, \pi^k]$ *is a* $\lambda^k(\eta)$*-link,* π^k *is* $\lambda^k(\eta)$*-free,* κ^k *is both* $\lambda^k(\eta)$*-free and* $\lambda^k(\xi)$*-free, and* κ^k *has* Σ *outcome along* $\lambda^k(\xi)$ *iff* κ^k *has* Σ *outcome along* $\lambda^k(\eta)$. *Then* $[\mu^k, \pi^k]$ *is a* $\lambda^k(\xi)$*-link and* π^k *is* $\lambda^k(\xi)$*-free.*

PROOF. Without loss of generality, we can assume that κ^k is the shortest node extending π^k that is both $\lambda^k(\xi)$-free and $\lambda^k(\eta)$-free. By symmetry, we can assume that there is no $\rho^k \in (\pi^k, \kappa^k]$ such that ρ^k is the last node of a $\lambda^k(\delta)$-link for some $\delta \in \{\eta, \xi\}$ that is not both a $\lambda^k(\eta)$-link and a $\lambda^k(\xi)$-link, as by Lemma 8.3.1 (Nesting), the longest such ρ^k would also be $\lambda^k(\delta)$-free as κ^k is $\lambda^k(\delta)$-free.

We obtain a contradiction under the assumption that π^k is not $\lambda^k(\xi)$-free. As π^k is not $\lambda^k(\xi)$-free and $\pi^k \subset \lambda^k(\xi)$, there must be a $\lambda^k(\xi)$-link $[\widetilde{\mu}^k, \widetilde{\pi}^k]$

that restrains π^k. As $\pi^k \subseteq \kappa^k$ which is $\lambda^k(\xi)$-free, we must have $\widetilde{\pi}^k \subseteq \kappa^k$; and without loss of generality using Lemma 8.3.1 (Nesting), we can assume that $\widetilde{\pi}^k$ is $\lambda^k(\xi)$-free. By Lemma 8.5.6 (Reflection) if $\widetilde{\pi}^k = \kappa^k$, and the maximality of the choice of π^k if $\widetilde{\pi}^k \subset \kappa^k$, $[\widetilde{\mu}^k, \widetilde{\pi}^k]$ must also be a $\lambda^k(\eta)$-link. But then π^k is not $\lambda^k(\eta)$-free, yielding the desired contradiction.

As π^k is $\lambda^k(\xi)$-free, it follows from the minimality of the choice of κ^k that $\pi^k = \kappa^k$, so by Lemma 8.5.6 (Reflection) that $[\mu^k, \pi^k]$ is a $\lambda^k(\xi)$-link. ⊣

The next lemma shows that if we have a $\lambda^r(\eta)$-link and a $\lambda^k(\eta^-)$-link, whose last nodes share a common antiderivative, and that antiderivative is switched at η, then their derived links on any tree are either disjoint, or the intersection of the derived links on lower level trees is generated by a primary $\lambda^r(\eta)$-link that is also a $\lambda^k(\eta^-)$-link.

LEMMA 8.5.8 (Alternating Link Lemma). *Fix $\eta \in T^0$, and let $\kappa^j = \lambda^j(\eta^-) \wedge \lambda^j(\eta)$ for all $j \leq \ell$. Suppose that there are $k, r \leq \ell$, $\widetilde{\mu}^k \in T^k$ and $\mu^r \in T^r$ such that $[\mu^r, \kappa^r]$ is a primary $\lambda^r(\eta)$-link and $[\widetilde{\mu}^k, \kappa^k]$ is a $\lambda^k(\eta^-)$-link. Let $[\mu^j, \pi^j]$ be the $\lambda^j(\eta)$-link derived from $[\mu^r, \kappa^r]$ for all $j \leq r$, and let $[\widetilde{\mu}^j, \widetilde{\pi}^j]$ be the $\lambda^j(\eta^-)$-link derived from $[\widetilde{\mu}^k, \kappa^k]$ for all $j \leq k$. Let $s = \min\{k, r\}$. Then:*

(i) *$k \neq r$.*
(ii) *If $i \leq s$, $[\mu^i, \pi^i) \cap [\widetilde{\mu}^i, \widetilde{\pi}^i) \neq \emptyset$ and $[\widetilde{\mu}^k, \kappa^k]$ is a primary $\lambda^k(\eta^-)$-link, then there is a $j < s$ such that $[\mu^j, \pi^j] \cap [\widetilde{\mu}^j, \widetilde{\pi}^j] = [\widehat{\mu}^j, \widehat{\pi}^j]$ is both a primary $\lambda^j(\eta^-)$-link and a primary $\lambda^j(\eta)$-link. Furthermore, $\widehat{\pi}^j$ is $\lambda^j(\eta^-)$-free if $k > r$, and is $\lambda^j(\eta)$-free if $r > k$.*

PROOF. It follows from Lemma 8.4.1(ii) (Free Extension) that for all $j < \ell$, $\mathrm{up}(\kappa^j) = \kappa^{j+1}$. We note that η switches κ^r, else $\lambda^r(\eta^-) = \kappa^r$ so $[\widetilde{\mu}^r, \kappa^r]$ would not be a $\lambda^r(\eta^-)$-link.

(i): We assume that $k = r$ and obtain a contradiction. As $[\mu^r, \kappa^r]$ is a primary $\lambda^r(\eta)$-link, κ^r has Π outcome along $\lambda^r(\eta)$, so has Σ outcome along $\lambda^r(\eta^-)$. Thus $[\widetilde{\mu}^r, \kappa^r]$ must be derived from a higher-level link, so $\kappa^{r+1} = \mathrm{up}(\kappa^r)$ must have Π outcome along $\lambda^{r+1}(\eta^-)$ and its principal derivative, κ^r, along $\lambda^r(\eta^-)$ must be its initial derivative along both $\lambda^r(\eta^-)$ and $\lambda^r(\eta)$. Thus $\mu^r = \kappa^r$, so $[\mu^r, \kappa^r]$ cannot be a $\lambda^r(\eta)$-link, contrary to hypothesis.

(ii): We first assume that $s = r < k$. Now κ^{r+1}, $\widetilde{\mu}^{r+1}$ and $\widetilde{\pi}^{r+1}$ are all derivatives of κ^{k+1} along $\lambda^{r+1}(\eta^-)$, so are comparable. Furthermore, by Lemma 8.5.1(ii,iii) (Freeness Transmission), both κ^{r+1} and $\widetilde{\pi}^{r+1}$ are $\lambda^{r+1}(\eta^-)$-free. So as $\widetilde{\mu}^{r+1}$ is an initial derivative of $\widetilde{\mu}^k$, $\widetilde{\pi}^{r+1}$ is the shortest $\lambda^{r+1}(\eta^-)$-free derivative of κ^k and $[\widetilde{\mu}^{r+1}, \widetilde{\pi}^{r+1}]$ is a $\lambda^{r+1}(\eta^-)$-link, we must have $\widetilde{\mu}^{r+1} \subseteq \kappa^{r+1}$ and either $\kappa^{r+1} = \widetilde{\pi}^{r+1}$ or $\widetilde{\pi}^{r+1} \subset \kappa^{r+1}$. In the latter case, it follows from Lemma 8.5.2 (Link Comparison) that $[\mu^i, \pi^i]$ and $[\widetilde{\mu}^i, \widetilde{\pi}^i]$ are disjoint for all $i \leq r$. Suppose that $\kappa^{r+1} = \widetilde{\pi}^{r+1}$. As the $\lambda^r(\eta)$-link $[\mu^r, \kappa^r]$ is primary, κ^{r+1} must have Σ outcome along $\lambda^{r+1}(\eta)$, so as η switches κ^r and hence κ^{r+1}, $\widetilde{\pi}^{r+1} = \kappa^{r+1}$ must have Π outcome along $\lambda^{r+1}(\eta^-)$ and so $\widetilde{\pi}^r = \mu^r \subset \pi^r = \kappa^r$;

therefore for all $i \leq r$, $\widetilde{\pi}^i$ is also the principal derivative of $\widetilde{\pi}^r$ along $\lambda^j(\eta)$. Thus for all $i \leq r$, $[\mu^i, \pi^i] \cap [\widetilde{\mu}^i, \widetilde{\pi}^i] = [\mu^i, \widetilde{\pi}^i]$. Now if $[\mu^i, \pi^i) \cap [\widetilde{\mu}^i, \widetilde{\pi}^i) \neq \emptyset$, then there is a greatest $j \leq r - 1$ such that $|[\mu^j, \widetilde{\pi}^j]| > 1$, and for this j, $\mathrm{up}(\mu^j) = \mu^{j+1} = \widetilde{\pi}^{j+1} = \mathrm{up}(\widetilde{\pi}^j)$, so $[\mu^j, \widetilde{\pi}^j]$ is both a primary $\lambda^j(\eta)$-link and a primary $\lambda^j(\eta^-)$-link. Furthermore, as $\widetilde{\pi}^j$ is the principal derivative of κ^k along $\lambda^j(\eta^-)$, it follows from Lemma 8.5.1(ii,iii) (Freeness Transmission) that $\widetilde{\pi}^j$ is $\lambda^j(\eta^-)$-free.

Now assume that $s = k \leq r$. By (i), $k < r$. The proof is symmetrical to that of the preceding paragraph, interchanging k and r, η and η^-, μ and $\widetilde{\mu}$, and π and $\widetilde{\pi}$. ⊣

As the construction proceeds, nodes may be switched many times. We will need to carefully track the effects of this switching process. The next lemma tells us that if a node is free and has a given outcome at η_0, and returns to the true path at a later stage η_1 as a free node, then all antiderivatives of the node have the same outcome along η_1 as they had along η_0.

LEMMA 8.5.9 (Outcome Restoration Lemma). *Fix $\delta, \eta_0, \eta_1 \in T^0$ such that $\eta_0 \subseteq \eta_1$. For all $j \leq \ell$, let $\delta^j = \mathrm{up}^j(\delta)$, assume that $\delta^j \subset \lambda^j(\eta_0)$, and let ρ^j be the immediate successor of δ^j along $\lambda^j(\eta_0)$. Assume that δ is η_i-free for $i \leq 1$, and that for all $j \leq \ell$, δ^j has Σ outcome along $\lambda^j(\eta_0)$ iff δ^j has Σ outcome along $\lambda^j(\eta_1)$. Then for all $j \leq \ell$, $\rho^j \subseteq \lambda^j(\eta_1)$.*

PROOF. As δ is η_i-free for $i \leq 1$, it follows from Lemma 8.5.1(i) (Freeness Transmission) that δ^j is $\lambda^j(\eta_i)$-free for all $j \leq \ell$ and $i \leq 1$.

We proceed by induction on $j \leq \ell$. If $j = 0$, then $\lambda^0(\eta_i) = \eta_i$ for $i \leq 1$. Thus by hypothesis, $\rho \subseteq \lambda^0(\eta_0) = \eta_0 \subseteq \eta_1 = \lambda^0(\eta_1)$, so the lemma follows in this case.

Suppose that $j > 0$. We consider two cases.

Case 1: δ^{j-1} has Π outcome ρ^{j-1} along $\lambda^{j-1}(\eta_0)$. By Definition 2.4.1, $\rho^j = \delta^j{}^\frown\langle\rho^{j-1}\rangle$. By induction, δ^{j-1} has Π outcome ρ^{j-1} along $\lambda^{j-1}(\eta_1)$. By (2.6.2), δ^{j-1} is the only derivative of δ^j along $\lambda^{j-1}(\eta_1)$ that has Π outcome along $\lambda^{j-1}(\eta_1)$, so by Definition 2.4.1, $\delta^j{}^\frown\langle\rho^{j-1}\rangle = \rho^j \subseteq \lambda^{j-1}(\eta_1)$ as desired.

Case 2: δ^{j-1} has Σ outcome ρ^{j-1} along $\lambda^{j-1}(\eta_0)$. Let $\tau^{j-1} = \mathrm{out}(\rho^j)$, and let $\sigma^{j-1} = (\tau^{j-1})^-$. As δ^{j-1} is $\lambda^{j-1}(\eta_0)$-free, it follows from the definition of links and (2.6.2) that all derivatives of δ^j must have Σ outcome along $\lambda^{j-1}(\eta_0)$, so σ^{j-1} is the initial derivative of δ^j along $\lambda^{j-1}(\eta_0)$. As δ^{j-1} is $\lambda^{j-1}(\eta_1)$-free and δ^{j-1} has Σ outcome ρ^{j-1} along $\lambda^{j-1}(\eta_1)$ by hypothesis, it follows from (2.6.2) and the definition of primary links that all derivatives of δ^j must have Σ outcome along $\lambda^{j-1}(\eta_1)$. Hence σ^{j-1} is the principal derivative of δ^j along $\lambda^{j-1}(\eta_1)$ and has Σ outcome along $\lambda^{j-1}(\eta_1)$. As σ^{j-1} is an initial derivative, $\sigma^{j-1} \subseteq \delta^{j-1}$, and so $\tau^{j-1} \subseteq \rho^{j-1} \subseteq \lambda^{j-1}(\eta_0)$; and as, by induction, $\rho^{j-1} \subseteq \lambda^{j-1}(\eta_1)$, we must have $\tau^{j-1} \subseteq \lambda^{j-1}(\eta_1)$. Thus $\rho^j = \delta^j{}^\frown\langle\tau^{j-1}\rangle \subseteq \lambda^j(\eta_1)$. ⊣

8.6. Framework Lemmas

This section is devoted to proving general lemmas about the operation of the framework. It analyzes sentence decomposition as well as various types of derived assignments.

8.6.1. Monotonic Sentence Decomposition. We begin by proving the Monotonic Sentence Decomposition Lemma introduced in Section 2.11.

LEMMA 8.6.1 (Monotonic Sentence Decomposition Lemma). *Suppose that a construction satisfies clauses (i)–(iii) of Definition 2.3.1 and clauses (i)–(iii) of Definition 2.7.7, and that $\Lambda^k \in [T^k]$ and $\eta^{k+1} \in T^{k+1}$ are given such that η^{k+1} is $\lambda(\Lambda^k)$-free. Then clause (ii) of Definition 2.10.1 is satisfied whenever a monotonic sentence decomposition process for directing sentences is followed, and clause (iii) of Definition 2.10.1 is satisfied whenever a monotonic sentence decomposition process for action sentences is followed, the construction has infinite support, and support propagates.*

PROOF. By Lemma 8.1.1(i,ii) (Limit Path), η^{k+1} has initial and principal derivatives μ^k and π^k, respectively, along Λ^k. Let F be the set of Λ^k-free derivatives of η^{k+1}. We first determine the nature of F in terms of the outcome of π^k along Λ^k. First suppose that π^k has Π outcome along Λ^k. If $\mu^k = \pi^k$, then by (2.6.2) and Lemma 8.4.2(i) (Free Derivative), $F = \{\pi^k\}$. Suppose that $\mu^k \neq \pi^k$. Then $[\mu^k, \pi^k]$ is a primary Λ^k-link, so by (2.6.2), no derivative of η^{k+1} other than π^k can be Λ^k-free. By Lemma 8.4.2(i) (Free Derivative), π^k is Λ^k-free, so $F = \{\pi^k\}$. Finally, if π^k has Σ outcome along Λ^k, then Lemma 8.4.2(ii) (Free Derivative) tells us that F is infinite.

Suppose that a monotonic sentence decomposition is followed. We prove only the case in which the decomposition bounds a block of existential quantifiers of the directing sentence S. A proof for the case in which the decomposition bounds a block of universal quantifiers of the sentence S is obtained from the proof below by interchanging even and odd, activated and validated, and replacing S with $\neg S$.

As the decomposition bounds a block of existential quantifiers of the sentence S, k must be even. By Definition 2.2.2, Π outcomes on T^k are validated outcomes, and Σ outcomes on T^k are activated outcomes.

First suppose that π^k has Π outcome along Λ^k, and so $F = \{\pi^k\}$. Then π^k is validated along Λ^k, so witnesses $\leq$ the respective parameters have been found for the existential quantifiers of S that were bounded in passing from T^{k+1} to T^k. It now follows that S is true. Furthermore, by Definition 2.2.2, η^{k+1} is validated along $\lambda(\Lambda^k)$. Clause (ii) of Definition 2.10.1 now follows in this case.

Now suppose that π^k has Σ outcome along Λ^k, and so F is infinite. By Remark 2.4.2(vi), all nodes in F have Σ outcome along Λ^k. For each $\nu^k \in F$, there is no set of witnesses for the existential quantifiers that have been bounded, each less than or equal to its associated bounding parameter. By monotonicity, there is no bound on the sizes of parameters as ν^k ranges over the Λ^k-free derivatives of η^{k+1}, so S is false. Furthermore, by Definition 2.2.2, η^{k+1} is activated along $\lambda(\Lambda^k)$. Clause (ii) of Definition 2.10.1 now follows in this case.

Clause (iii) of Definition 2.10.1 follows as above from the monotonic sentence decomposition and the hypotheses that each Λ^k-free node has infinite support and that support propagates. ⊣

8.6.2. Derived Assignments. Several derived assignments were introduced. The one that was used most often was the standard derived assignment introduced in Definition 2.7.8, and we proved, in Lemma 2.12.2 (Standard Derived Assignment) that this assignment satisfied clauses (i)– (iii) of Definition 2.7.7. We now show that these clauses are satisfied as well for the preferential derived assignment

LEMMA 8.6.2 (Preferential Derived Assignment Lemma). *The preferential derived assignment satisfies clauses* (*i*)–(*iii*) *of Definition* 2.7.7.

PROOF. We proceed by induction on $n - k$, assuming that the construction has level n. Clause (i) of Definition 2.7.7 is immediate from the definition of the preferential derived assignment, and clause (ii) of definition 2.7.7 has a proof identical to that given in Lemma 2.12.2 for the standard derived assignment.

It remains to show that the preferential derived assignment satisfies clause (iii) of Definition 2.7.7. Let $\mathcal{R}_0$ and $\mathcal{R}_1$ partition the requirements, with $\mathcal{R}_0$ the preferred class of requirements. It follows easily from the definition of preferred derived assignments that each subblock along a path through T^k is finite. As each requirement in $\mathcal{R}_1$ is assigned to at most one node along any path through a block, a proof virtually identical to that of Lemma 8.1.1(iv) (Limit Path) shows that there are only finitely many subblocks within a block. Clause (iii) of Definition 2.7.7 now follows for the preferential derived assignment. (Note that the proof of Lemma 8.1.1(iv) (Limit Path) does not appeal to clause (iii) of Definition 2.7.7, but does appeal to clause (ii) of that definition; as clause (ii) has been verified for this derived assignment, there is no circularity in this proof.) ⊣

Both the standard derived assignment and the preferential derived assignment work well when basic modules have a single non-terminal node; but when basic modules are larger, we need a careful transmission of these modules to lower levels, ensuring that successive nodes of the module have successive derivatives, whenever possible.

DEFINITION 8.6.3. Let $\eta^k, \nu^k \in T^k$ be given. We say that the pair $\langle \eta^k, \nu^k \rangle$ is *connected* if $(\nu^k)^- = \eta^k$, and we say that this pair is *module-connected* if it is connected and both nodes are antiderivatives of a connected pair of nodes lying in the same realization of some basic module.

There are three basic situations in which we can expect a module-connected pair $\langle \eta^k, \nu^k \rangle$ on T^k to give rise to a module-connected pair $\langle \eta^j, \nu^j \rangle$ on T^j for $j < k$ under a derived assignment. The first is when η^j is the principal derivative of η^k along $\text{out}^j(\nu^k)$. The second is when both η^k and ν^k have level $> j$ so we may assume that $j = k - 1$, η^k has Π outcome along ν^k so can be expected to have infinitely many derivatives along a path through T^j computing a path containing ν^k, ν^k lies along the path computed by ν^j, and ν^k has not yet specified its final derivative along the path. The third is when j is the first level producing more than a single derivative of η^k along a given path, and the derivative of η^k along a path through T^{j+1} containing a derivative of ν^k has Π outcome along the path. We specify these conditions below, and call a derived assignment *module-respecting* if it gives rise to module-connected pairs in these three cases.

DEFINITION 8.6.4. A derived assignment is said to be *module-respecting* if the following conditions hold whenever $j < k$, $\eta^j, \nu^j \in T^j$ and $\eta^k, \nu^k \in T^k$:

(i) If $j = k - 1$, $\langle \eta^k, \nu^k \rangle$ is module-connected, $\langle \eta^j, \nu^j \rangle$ is connected, and η^j is the principal derivative of η^k along ν^k, then $\text{up}(\eta^j) = \eta^k$.

(ii) If $j = k - 1$, $\langle \eta^k, \nu^k \rangle$ is module-connected, $\langle \eta^j, \nu^j \rangle$ is connected, $\text{lev}(\eta^k) \geq k$, η^k has Π outcome along ν^k, η^j has Σ outcome along ν^j, $\text{up}(\eta^j) = \eta^k$, and there is no $\xi^j \subset \eta^j$ such that $\text{up}(\xi^j) = \nu^k$ and ξ^j has Π outcome along ν^j, then $\text{up}(\nu^j) = \nu^k$.

(iii) If $\langle \eta^k, \nu^k \rangle$ is module-connected, $\langle \eta^j, \nu^j \rangle$ is connected, $\text{lev}(\eta^j) = j + 1$, $\text{lev}(\nu^k) \geq j + 1$, $\text{up}^k(\nu^j) = \nu^k$, and $(\text{out}^{j+1}(\nu^k))^-$ has Π outcome along $\text{out}^{j+1}(\nu^k)$, then $\text{up}^k(\eta^j) = \eta^k$.

We now show that such derived assignments exist.

LEMMA 8.6.5 (Module-Respecting Derived Assignment Lemma). *If all derived assignments on T^i for $i > j$ are module-respecting, then there is a module-respecting derived assignment for T^j satisfying conditions (i)–(iii) of Definition* 2.7.7.

PROOF. We follow the pattern used to define the standard derived assignment, but insist on choosing antiderivatives to satisfy clauses (i)–(iii) of Definition 8.6.4 when such choices are required in order to have a module-respecting derived assignment. If one of these clauses specifies an antiderivative that does not satisfy clause (i) of Definition 2.7.7, then we halt further assignments along that path. Thus the assignment may only be partial; however, as in the proof of Lemma 2.12.2 (Standard Derived Assignment), we see that this partial assignment satisfies clauses (i)–(iii) of Definition 2.7.7, and clauses (i)–(iii) of Definition 8.6.4.

The proof that this assignment is total depends on the ability to show that in the situations described above, we are able to choose the appropriate derivatives satisfying (2.6.1)–(2.6.4). Of these conditions, all follow easily except for (2.6.4). This latter condition follows from Lemma 8.4.1(i) (Free Extension) for clauses (i) and (ii) of Definition 8.6.4, and from Lemma 8.4.1(i) (Free Extension) and Lemma 8.4.2(i) (Free Derivative) for clause (iii). (Note that we may use these lemmas without fear of circularity, as clauses (i)–(iii) of Definition 2.7.7 have been verified.) ⊣

The next lemma is used in the proof of the Minimal Pair Theorem.

LEMMA 8.6.6. *Suppose that the preferential derived assignment is followed on* T^0*, and let* $\mathcal{R}_0$ *be the preferred set of requirements and let* $\mathcal{R}_1$ *be its complement. Let* $\eta_i \subset \nu_i \in T^0$ *for* $i \leq 1$ *be given such that* η_i *is validated along* ν_i, $(\nu_i)^- = \eta_i$, $\nu_0 \subset \nu_1$, $\mathrm{up}^2(\eta_0) = \mathrm{up}^2(\eta_1)$, *and there is no* $\xi \in (\eta_0, \eta_1)$ *such that* $\mathrm{up}^2(\xi) = \mathrm{up}^2(\eta_0)$ *and* ξ *is validated along* η_1. *Then there is at most one* $\sigma \in [\nu_0, \nu_1)$ *such that a requirement in* $\mathcal{R}_1$ *is assigned to* σ, $\mathrm{wt}(\mathrm{up}(\sigma)) \leq \mathrm{wt}(\mathrm{up}(\eta_0))$, *and* σ *is validated along* ν_1.

PROOF. Fix $i \in \{0, 1\}$. Suppose that a requirement in $\mathcal{R}_1$ is assigned to σ_i, $\mathrm{wt}(\mathrm{up}(\sigma_i)) \leq \mathrm{wt}(\mathrm{up}(\eta_0))$, $\sigma_i \in [\nu_0, \nu_1)$, σ_i is validated along η_1, and $\sigma_0 \subset \sigma_1$, in order to obtain a contradiction. For $j \in \{1, 2\}$, let $\sigma_i^j = \mathrm{up}^j(\sigma_i)$, $\eta_i^j = \mathrm{up}^j(\eta_i)$, and $\nu_i^j = \lambda^j(\nu_i)$. As σ_i has Π outcome along ν_i, Definition 2.4.1 implies that $(\nu_i^1)^- = \eta_i^1$. Also, let $\widetilde{\eta}^1$ be the initial derivative of $\eta^2 = \mathrm{up}^2(\eta_0) = \mathrm{up}^2(\eta_1)$ along η_0^1.

By (2.8.1) and (2.8.4), we must have $\sigma_i^1 \subset \eta_0^1$. By Lemma 8.1.1(ii) (Limit Path) and (2.8.4), we must have $\sigma_1^1 \subset \sigma_0^1$. Now $\sigma_i^1, \widetilde{\eta}^1 \subseteq \eta_0^1$, so σ_i^1 and $\widetilde{\eta}^1$ must be comparable; and as a requirement in $\mathcal{R}_1$ is assigned to σ_i^1 and a requirement in $\mathcal{R}_0$ is assigned to $\widetilde{\eta}^1$, we must have $\sigma_i^1 \neq \widetilde{\eta}^1$. We cannot have $\sigma_i^1 \subset \widetilde{\eta}^1$, as Definition 2.4.1 would preclude the existence of $\eta_0 \subset \sigma_i$ if σ_i^1 has Σ outcome along $\widetilde{\eta}^1$, and Lemma 8.1.1(i) (Limit Path) and (2.6.2) would preclude the existence of $\sigma_i \subset \eta_1$ if σ_i^1 has Π outcome along $\widetilde{\eta}^1$. We thus conclude that $\widetilde{\eta}^1 \subset \sigma_i^1$.

We now consider the situation on T^2. An $\widetilde{\eta}^1 \subset \sigma_i^1 \subset \eta_0^1$, it follows from clause (ii) of Remark 2.4.2 that $\eta^2 = \mathrm{up}(\widetilde{\eta}^1) = \mathrm{up}(\eta_0^1)$ is comparable with σ_i^2. We cannot have $\sigma_i^2 \subset \eta^2$, as σ_i^1 must have Π outcome along η_0^1, so as $\widetilde{\eta}^1 \subset \sigma_i^1 \subset \eta_0^1$, σ_i^2 must have Π outcome along $\lambda(\widetilde{\eta}^1) \subseteq \eta^2$ and Σ outcome along $\lambda(\eta_i^1) \subseteq \eta^2$, a contradiction; as a requirement in $\mathcal{R}_1$ is assigned to σ_i^2 and a requirement in $\mathcal{R}_0$ is assigned to η^2, we conclude that $\eta^2 \subset \sigma_i^2$. Furthermore, as $\widetilde{\eta}^1 \subset \sigma_i^1 \subset \eta_0^1$, it follows from (2.6.2) that η^2 must have Π outcome along σ_i^2.

Fix $\tau_0 \subseteq \eta_1$ such that $\tau_0^- = \sigma_0$ and let $\tau_0^1 = \lambda(\tau_0)$. As σ_0 is validated along η_1, σ_0 has Π outcome along τ_0, so by Lemma 8.1.1(ii) (Limit Path), setting $\tau_0^1 = \lambda(\tau_0)$, we must have $(\tau_0^1)^- = \sigma_0^1$. As the preferential derived assignment is implemented on T^1, given any path Γ^1 through T^1 such that $\tau_0^1 \subset \Gamma^1$, there will be a segment of Γ^1 beginning a subblock at τ_0^1 such that the nodes in that segment will be antiderivatives of those nodes $\xi^2 \subset \sigma_0^2$ having Π outcome along σ_0^2 to which requirements in $\mathcal{R}_0$ are assigned, and antiderivatives will be chosen in strict order of length until either one of the nodes in the segment has Π outcome along Γ^1 or all possible antiderivatives with this property are exhausted. Note that η^2 lies in this set of antiderivatives. Furthermore, as the preferential derived assignment is implemented on T^0, no node in $(\eta_0^1, \sigma_0^1]$ can be switched until we reach a node $\rho \subset \Lambda^0$ that is a derivative of a node in such a subblock whose antiderivative is η^2 and which has Σ outcome along $\lambda(\rho)$. By definition, we must have $\rho = \eta_1$, so in particular, $\eta_1 \subset \sigma^1$ yielding the desired contradiction. $\dashv$

8.7. Level Analysis Lemmas

This section is devoted to proving the lemmas about the operation of the framework at the lower levels.

8.7.1. The Σ_1 Level. We begin at level 1 with the Σ_1-Analysis Lemma, also listed as Lemma 3.1.8. We note that the proof of this lemma applies to T^k and T^{k+1} in place of T^0 and T^1, so the lemma relativizes.

LEMMA 8.7.1 (Σ_1 Analysis Lemma). *Let $\eta \subset \nu \subseteq \rho \subset \tau \in T^0 \cup [T^0]$ be given such that $\nu^- = \eta$ and η is τ-free. Then the following conditions hold*:

(*i*) *If $|\tau| = \infty$, then $|\{\xi \in (\eta, \tau) : \eta$ is ξ-free$\}| = \infty$.*

(*ii*) *If η is ρ-free, then* $\mathrm{up}(\eta)$ *is $\lambda(\rho)$-free.*

(*iii*) $\mathrm{up}(\eta) \subset \lambda(\nu), \lambda(\rho)$.

(*iv*) *If $\eta^1 \subset \Lambda^1 \in [T^1]$, $\Lambda^0 \in [T^0]$, $\lambda(\Lambda^0) = \Lambda^1$, and η^1 is Λ^1-free and has Π outcome along Λ^1, then η^1 has infinitely many Λ^0-free derivatives.*

(*v*) *If $\eta^1 \subset \Lambda^1 \in [T^1]$, $\Lambda^0 \in [T^0]$, $\lambda(\Lambda^0) = \Lambda^1$, and η^1 is Λ^1-free and has Σ outcome along Λ^1, then η^1 has a unique Λ^0-free derivative π with Π outcome along Λ^0, and π is the principal derivative of η^1 along Λ^0.*

(*vi*) *η is ν-free, and if η is the principal derivative of* $\mathrm{up}(\eta)$ *along ν, then* $(\lambda(\nu))^- = \mathrm{up}(\eta)$.

PROOF. (i): Suppose that the construction is a level n construction, and let $\Lambda^j = \lambda^j(\tau)$ for all $j \leq n$. By Lemma 8.4.2(i) (Free Derivative), if $\xi^n \subset \Lambda^n$, then the principal derivative ξ of ξ^n along Λ^0 is Λ^0-free; hence if $|\Lambda^n| = \infty$, then there are infinitely many Λ^0-free $\xi \subset \Lambda^0$. If $|\Lambda^n| < \infty$, then there must be a longest $\xi^n \subset \Lambda^n$ such that $\mathrm{lev}(\xi^n) = n$ and ξ^n has Π outcome along τ^n; ξ^n will have infinitely many derivatives along Λ^{n-1} all of which will be Λ^{n-1}-free, so applying Lemma 8.4.2(i) (Free Derivative) to each of these derivatives as above, we again see that there are infinitely many Λ^0-free $\xi \subset \Lambda^0$. By Lemma 8.3.7(ii) (Free and True), each immediate successor $\nu \subset \Lambda^0$ of a Λ^0-free ξ

is Λ^0-true. Hence by Lemma 8.3.6 (Link Preservation) and Lemma 8.3.7(ii) (Free and True), for such $\xi = \nu^-$, η is ξ-free iff η is Λ^0-free, proving (i).

(ii): This is the statement of Lemma 8.3.7(iii) (Free and True).

(iii): By Lemma 8.4.1(i) (Free Extension), $\mathrm{up}(\eta) \subset \lambda(\nu)$. As η is τ-free, it follows from Lemma 8.3.3(i) (Link Analysis) that $\mathrm{up}(\eta) \subset \lambda(\tau)$. (iii) now follows from Remark 2.4.2(ii).

(iv): Immediate from Lemma 8.4.2(ii) (Free Derivative).

(v): By Lemma 8.1.1(ii) (Limit Path), η^1 has a principal derivative π along Λ^0, and by Definition 2.4.1, π has Π outcome along Λ^0. By (2.6.2), π is the only derivative of η^1 having Π outcome along Λ^0. By Lemma 8.4.2(i) (Free Derivative), π is Λ^0-free.

(vi): Immediate from Lemma 8.1.1(ii) (Limit Path). ⊣

8.7.2. The Δ_2 Level. We now prove a lemma identifying properties first encountered at the Δ_2 level.

LEMMA 8.7.2 (Δ_2 Analysis Lemma). *For a Δ_2 construction, fix $\eta^2 \in T^2$ such that* $\mathrm{lev}(\eta^2) = 2$. *Suppose that $\xi^1, \eta^1 \in T^1$ and* $\mathrm{up}(\xi^1) = \mathrm{up}(\eta^1) = \eta^2$. *Then*:

(*i*) *ξ^1 and η^1 are comparable nodes of T^1, say $\xi^1 \subset \eta^1$, and all nodes in $[\xi^1, \eta^1]$ are derived from η^2.*

(*ii*) *If $\eta^1 \subset \lambda(\eta)$ is the longest derivative of η^2 that is $\subset \lambda(\eta)$, then either η^1 has Π outcome along $\lambda(\eta)$ and is the only $\lambda(\eta)$-free derivative of η^2, or η^1 has Σ outcome along $\lambda(\eta)$ and all derivatives of η^2 along $\lambda(\eta)$ are $\lambda(\eta)$-free.*

PROOF. (i): Let $\gamma^1 = \mathrm{out}(\eta^2)$. If $\sigma^2 \subset \eta^2$, then by Lemma 8.1.1(ii) (Limit Path), σ^2 has a principal derivative $\sigma^1 \subset \gamma^1$, and σ^1 is the principal derivative of σ^2 along any $\rho^1 \supseteq \gamma^1$. Furthermore, if $\mathrm{lev}(\sigma^2) = 2$, then as we have a Δ_2 construction, σ^2 has Σ outcome along η^2, so by Definition 2.4.1, σ^1 must have Π outcome along γ^1. By (2.6.2) if $\mathrm{lev}(\sigma^2) = 2$ and by (2.6.3) if $\mathrm{lev}(\sigma^2) = 1$, σ^1 is the longest derivative of σ^2 along any $\rho^1 \supseteq \gamma^1$. Furthermore, if $\gamma = \mathrm{out}(\gamma^1)$, $\eta \subset \Lambda^0$ and $\mathrm{up}(\eta) = \eta^1$, then by Remark 2.4.2(ii), no $\rho \in (\gamma, \eta)$ can switch any node $\subset \xi^1$, else η^2 would not be on the current path through T^2 at η. It now follows that $\lambda^2(\rho) \supseteq \eta^2$ for all $\rho \in (\gamma, \eta)$, and that for all $\rho^1 \in [\gamma^1, \lambda(\eta)]$, η^2 is the shortest ρ^1-consistent node.

As $\lambda(\gamma^1) = \eta^2$, η^2 will be the only γ^1-consistent node, so $\mathrm{up}(\gamma^1) = \eta^2$. If $\nu \in (\gamma, \eta]$, $\mathrm{up}^2(\nu^-) = \eta^2$, and $\mathrm{up}(\nu^-)$ has Σ outcome along $\lambda(\nu)$, then η^2 will have Π outcome along $\lambda^2(\nu)$, so as this is a Δ_2 construction, it will again be the case that η^2 is the only $\lambda(\nu)$-consistent node, and $(\lambda(\nu))^- = \mathrm{up}(\nu^-)$. Hence $(\mathrm{up}(\nu))^- = \mathrm{up}(\nu^-)$, and $\mathrm{up}^2(\nu) = \eta^2$. Furthermore, if $\mathrm{up}(\nu^-)$ has Π outcome along $\lambda(\nu) = \mathrm{up}(\nu)$, then by (2.6.2), no extension of $\mathrm{up}(\nu)$ can be a derivative of η^2. We conclude that the derivatives of η^2 on T^1 form an interval of T^1, with each non-initial derivative of η^2 on T^1 extending the Σ outcome of its predecessor.

(ii): By the proof of (i), if $\eta^2 \subset \lambda^2(\eta)$, then $\lambda(\eta)$ can be partitioned into intervals as follows: the first interval consists of the derivatives of nodes $\subset \eta^2$

or incomparable with η^2; the second interval consists of derivatives of η^2; and the third interval consists of derivatives of proper extensions of η^2. As this is a level 2 construction, all links on T^1 are primary links. So a derivative of η^2 can only be restrained by a primary $\lambda(\eta)$-link $[\mu^1, \pi^1]$ if $\text{up}(\mu^1) = \text{up}(\pi^1) = \eta^2$. If η^1 has Π outcome along $\lambda(\eta)$, then $\eta^1 = \pi^1$ so is not restrained by this link; and if η^1 has Σ outcome along $\lambda(\eta)$, then no such link exists, so all derivatives of η^2 along $\lambda(\eta)$ are $\lambda(\eta)$-free. ⊣

8.7.3. The Δ_3 Level. Δ_3 constructions are constructions all of whose requirements are at a level ≤ 3, and such that the Π outcome of a level 3 node produces a terminal node on T^3. We present some of the properties of Δ_3 constructions. The first lemma shows that the uniqueness of initial derivatives carries up to the Δ_3 level, and is a restatement of Lemma 6.1.1.

LEMMA 8.7.3. *Let a Δ_3 construction be given with true path $\Lambda^0 \in T^0$. Suppose that $\eta^3 = \lambda^3(\sigma)$ for some $\sigma \subset \Lambda^0$. Then:*

(*i*) *There is a unique $\xi^2 \in \{\lambda^2(\xi) : \xi \subset \Lambda^0\}$ such that ξ^2 is an initial derivative of η^3.*

(*ii*) *There is a unique $\xi^1 \in \{\lambda(\xi) : \xi \subset \Lambda^0\}$ such that ξ^1 is an initial derivative of η^3.*

PROOF. (i): Suppose that $\xi^2 \neq \mu^2 \in T^2$ and both ξ^2 and μ^2 are initial derivatives of η^3, in order to obtain a contradiction. Then $\xi^2|\mu^2$; let $\kappa^2 = \xi^2 \wedge \mu^2$. By (2.6.1) and Remark 2.4.2(i), $\text{out}(\eta^3) \subseteq \kappa^2$, so by Lemma 8.1.2 (Out), $\lambda(\text{out}(\eta^3)) = \eta^3$; and by Lemma 8.1.1(i) (Limit Path), $\lambda(\xi^2) = \lambda(\mu^2) = \eta^3$. Thus by Definition 2.4.1, all nodes in $[\kappa^2, \xi^2)$ must have Σ outcome along ξ^2, all nodes in $[\kappa^2, \mu^2)$ must have Σ outcome along μ^2, and κ^2 cannot be an initial derivative of $\kappa^3 = \text{up}(\kappa^2)$. It now follows that $\text{lev}(\kappa^3) = 3$ and κ^3 has Π outcome along η^3. But then as we have a Δ_3 construction, no requirement is assigned to η^3, so η^3 has no derivatives on T^2, yielding a contradiction.

(ii) Suppose that $\xi^1 \neq \mu^1 \in T^1$ and both ξ^1 and μ^1 are initial derivatives of η^3, in order to obtain a contradiction. By (i), both ξ^1 and μ^1 are initial derivatives of the unique initial derivative ξ^2 of η^3 on T^2. By definition of initial derivatives and as $\xi^1 \neq \mu^1$, we must have $\xi^1|\mu^1$; let $\kappa^1 = \xi^1 \wedge \mu^1$. By (2.6.1) and Remark 2.4.2(i), $\text{out}(\xi^2) \subseteq \kappa^1$, so by Lemma 8.1.2 (Out), $\lambda(\text{out}(\xi^2)) = \xi^2$; and by Lemma 8.1.1(i) (Limit Path), $\lambda(\xi^1) = \lambda(\mu^1) = \xi^2$. Thus by Definition 2.4.1, all nodes in $[\kappa^1, \xi^1)$ must have Σ outcome along ξ^1, all nodes in $[\kappa^1, \mu^1)$ must have Σ outcome along μ^1. By (2.6.2), κ^1 has at most one derivative with Π outcome along Λ^0, so by Definition 2.4.1, κ^1 must have the same outcome along both μ^1 and ξ^1, contrary to the definition of κ^1. ⊣

The following Lemma restates Lemma 6.1.2.

LEMMA 8.7.4. *Let a Δ_3 construction be given. Fix $\eta^3 \in T^3$ and $\xi^2, \eta^2, \kappa^2 \in T^2$ such that $\text{up}(\xi^2) = \text{up}(\eta^2) = \eta^3$ and $\kappa^2 \subseteq \xi^2 \wedge \eta^2$. Then:*

(*i*) *If $\text{up}(\kappa^2) = \eta^3$, then $\text{up}(\sigma^2) = \eta^3$ for all $\sigma^2 \in [\kappa^2, \eta^2]$.*

(*ii*) *If* $\kappa^2 = \xi^2 \wedge \eta^2$, *then* $\mathrm{up}(\kappa^2) = \eta^3$.

PROOF. (i): By (2.6.2), all derivatives of η^3 along η^2 must have Σ outcome along η^2, and so by Lemma 8.1.1(i) (Limit Path), η^3 cannot have Σ outcome along $\lambda(\kappa^2)$, nor can it have Σ outcome along $\lambda(\eta^2)$. As this is a Δ_3 construction, it follows from Remark 2.4.2(ii) that $\lambda(\sigma^2) = \eta^3$ for all $\sigma^2 \in [\kappa^2, \eta^2]$. By (2.6.1)–(2.6.3) and as this is a Δ_3 construction, each such σ^2 must be a derivative of η^3.

(ii): Suppose that $\kappa^2 = \xi^2 \wedge \eta^2$. Now by Lemma 8.7.3, η^3 has a unique initial derivative $\mu^2 \in T^2$, and by Lemma 8.1.1(i) (Limit Path), $\mu^2 \subset \xi^2, \eta^2$; hence $\mu^2 \subseteq \kappa^2$. (ii) now follows from (i). ⊣

The following Lemma restates Lemma 6.1.3.

LEMMA 8.7.5. *Let a* Δ_3 *construction be given. Fix* $\eta^3 \in T^3$ *with initial derivative* $\eta^1 \subset \tau^1 \in \{\lambda(\sigma) : \sigma \subset \Lambda^0\}$ *such that* $\mathrm{up}^3(\eta^1) = \mathrm{up}^3(\tau^1) = \eta^3$. *Then there is a number k and* $\mu_0^1 \subseteq \pi_0^1 \subset \mu_1^1 \subseteq \pi_1^1 \subset \cdots \subset \mu_k^1 \subseteq \pi_k^1$ *such that all* $\xi^1 \in (\pi_i^1, \mu_{i+1}^1)$ *have* Σ *outcome along* μ_{i+1}^1 *for all* $i < k$, $\mathrm{up}(\mu_i^1) = \mathrm{up}(\pi_i^1)$ *for all* $i \leq k$, $(\mathrm{up}(\mu_{i+1}^1))^- = \mathrm{up}(\mu_i^1)$ *for all* $i < k$, π_i^1 *has* Π *outcome along* μ_{i+1}^1 *for all* $i < k$, $\mu_0^1 = \eta^1$ *and* $\pi_k^1 = \tau^1$. ⊣

PROOF. As η^1 is an initial derivative of η^3, it follows from Lemma 8.7.3(i) that $\eta^2 = \mathrm{up}(\eta^1)$ is also an initial derivative of η^3, and that $\eta^2 \subseteq \tau^2 = \mathrm{up}(\tau^1)$. By Lemma 8.7.4, if $\{\delta_i^2 : i \leq k\}$ is a listing of the nodes in $[\eta^2, \tau^2]$ in order of length, then $\mathrm{up}(\delta_i^2) = \eta^3$ for all $i \leq k$. Furthermore, by (2.6.2), for $i < k$, η_i^2 has Σ outcome along η_{i+1}^2. Let μ_i^1 be the initial derivative of δ_i^2 along τ^1 for all $i \leq k$, let π_i^k be the principal derivative of δ_i^2 along τ^1 for all $i < k$, and let $\pi_k^1 = \tau^1$. All the conclusions of the lemma except for the one relating to intervals of the form (π_i^1, μ_{i+1}^1) follow easily.

Suppose that $\sigma^1 \in (\pi_i^1, \mu_{i+1}^1)$. Then σ^1 must be a derivative of a node $\rho^2 \subseteq \delta_i^2$. If $\rho^2 \subset \delta_i^2$ has level 1 or has Σ outcome along μ_{i+1}^1, then by (2.6.2), (2.6.3) and Lemma 8.1.1(ii) (Limit Path), the principal (and longest) derivative of ρ^2 along τ^1 is $\subseteq \pi_i^1$. If ρ^2 has level 2 or level 3 and has Π outcome along μ_{i+1}^1, then σ^1 must have Σ outcome along μ_{i+1}^1. Thus the lemma holds. ⊣

The following lemma was stated as Lemma 6.1.3.

LEMMA 8.7.6. *Let a* Δ_3 *construction be given. Suppose that* $\mu^1 \subset \eta^1 \subset \pi^1 \subset \rho^1 \in T^1$, $[\mu^1, \pi^1]$ *is a primary* ρ^1*-link, and* η^1 *has* Π *outcome along* π^1. *Then* $\mathrm{wt}(\mathrm{up}(\pi^1)) < \mathrm{wt}(\mathrm{up}(\eta^1))$. *If, furthermore,* $\rho^1 \subset \Lambda^1 \in [T^1]$, *then* $\mathrm{up}^3(\eta^1) \not\subset \lambda^3(\Lambda^1)$.

PROOF. As $[\mu^1, \pi^1]$ is a primary ρ^1-link, $\mathrm{up}(\mu^1) = \mathrm{up}(\pi^1)$; set $\mu^2 = \mathrm{up}(\mu^1)$. Now by (2.6.2), $\mu^2 \subseteq \lambda(\mu^1), \lambda(\pi^1)$, so by Remark 2.4.2(ii), $\mu^2 \subseteq \lambda(\eta^1)$. Again by (2.6.2), $\eta^2 = \mathrm{up}(\eta^1) \subseteq \lambda(\eta^1)$; hence μ^2 and η^2 are comparable. Let $\widehat{\eta}^1$ be the initial derivative of η^2 along η^2. By Lemma 8.3.1 (Nesting) and

(2.6.2), we must have $\mu^1 \subset \widehat{\eta}^1$. Lemma 8.1.1(i) (Limit Path) now tells us that $\lambda(\mu^1) = \mu^2$ and $\lambda(\widehat{\eta}^1) = \eta^2$, and as a result, that $\mu^2 \subset \eta^2$. Hence by (2.8.1), $\text{wt}(\text{up}(\pi^1)) = \text{wt}(\pi^2) < \text{wt}(\eta^2) = \text{wt}(\text{up}(\eta^1))$.

Now suppose that $\rho^1 \subset \Lambda^1$, and let $\Lambda^i = \lambda^i(\Lambda^1)$ for $i = 2, 3$ and $\eta^3 = \text{up}^3(\eta^1)$. Suppose that $\eta^3 \subset \Lambda^3$ in order to obtain a contradiction. By Lemma 8.1.1(i) (Limit Path), η^3 has a unique initial derivative $\widetilde{\eta}^2 \subset \eta^2$; hence as $\mu^2 \subset \eta^2$, μ^2 and $\widetilde{\eta}^2$ are comparable. If $\mu^2 \subset \widetilde{\eta}^2$, then by Lemma 8.1.1(i) (Limit Path), $[\mu^1, \pi^1]$ would be a primary Λ^1-link properly restraining the initial derivative of $\widetilde{\eta}^2$ along Λ^1, so by Lemma 8.3.3(iii) (Link Analysis), $\widetilde{\eta}^2 \not\subset \Lambda^2$; but then it would follow from Lemma 8.1.1(i) (Limit Path) and Lemma 8.7.3 that $\eta^3 \not\subset \Lambda^3$, a contradiction.

We conclude that $\widetilde{\eta}^2 \subseteq \mu^2 \subset \eta^2$. Thus by Lemma 8.7.4, $\text{up}(\mu^2) = \eta^3$. Now μ^1 is the principal derivative of μ^2 along $\lambda(\eta^1)$, so μ^2 must have Π outcome along $\eta^2 \subseteq \lambda(\eta^1)$. But as $\text{up}(\mu^2) = \text{up}(\eta^2)$, we have contradicted (2.6.2). Thus we can conclude that $\eta^3 \not\subset \Lambda^3$. ⊣

8.8. Restraint Lemmas

Restraint Lemmas are lemmas that show that computations are preserved. As the level of the construction increases, the analysis of what happens at the higher levels is more complex, and this complexity carries over to the statements and proofs of the restraint lemmas. We collect the lemmas of this kind that do not involve backtracking in this section, beginning with the Σ_1 level.

LEMMA 8.8.1 (Restraint Lemma). *Suppose that for a given faithfully executed construction, action for η is supported at ρ only when η is ρ-free, sentence decomposition is monotonic, the construction admits consistent action, and clauses (ii) and (iii) of Definition* 3.1.7 *are satisfied. Then*:

(*i*) *If clause (i) of Definition* 3.1.7 *holds, a restraint condition $x \notin A^s$ is declared as action for η along ρ, and* $\text{up}(\eta) \subset \lambda(\rho)$, *then $x \notin A^{\text{wt}(\rho)}$*.

(*ii*) *Shared set action is compatible.*

(*iii*) *If clauses (i) and (iv) of Definition* 3.1.7 *holds with the inequality in (ia) changed to equality and clause (iia) of Definition* 3.1.7 *is replaced with* $\text{wt}(\text{up}(\eta)) = x$, *then shared functional action is compatible.*

PROOF. (i): Fix $\nu \subseteq \rho$ such that $\nu^- = \eta$. (The idea here is that, as the validated outcome of $\text{up}(\eta)$ has not been switched off the current path before ρ is reached, all numbers entering A through action for nodes in $[\nu, \rho)$ are greater than the use of the axiom that was newly declared at ν.) Assume the hypothesis of (i). By clauses (ia,d) of Definition 3.1.7, $x < u \leq \text{wt}(\lambda(\nu))$ and η is the principal derivative of $\eta^1 = \text{up}(\eta)$ along ν. By Lemma 8.1.1(ii) (Limit Path) and Remark 2.4.2(ii), $\nu^1 = \text{up}(\nu) = \lambda(\nu) \subseteq \lambda(\zeta)$ for all $\zeta \in [\nu, \rho]$.

Suppose that $\zeta \in [\nu, \rho]$ and action for σ supported at ζ specifies that $x \in A$, in order to obtain a contradiction. By hypothesis, σ must be ζ-free, hence by Lemma 8.3.3(i) (Link Analysis), $\sigma^1 = \text{up}(\sigma) \subseteq \lambda(\zeta)$ and hence σ^1 and ν^1 are comparable. If $\nu^1 = \sigma^1$, then by clause (iia) of Definition 3.1.7, $\text{wt}(\sigma^1) \leq x < u \leq \text{wt}(\lambda(\nu)) = \text{wt}(\nu^1) = \text{wt}(\sigma^1)$, a contradiction. If $\nu^1 \subset \sigma^1$, then by Definition 2.4.1 and Lemma 8.1.1(i) (Limit Path), σ^1 cannot have an initial derivative $\subset \nu$, so by (2.8.4), Lemma 8.1.1(i) (Limit Path), and clause (iia) of Definition 3.1.7, $u \leq \text{wt}(\lambda(\nu)) \leq \text{wt}(\sigma^1) \leq x$, a contradiction. We conclude that $\sigma^1 \subset \nu^1$. By clause (iiic) of Definition 3.1.7, $x \notin A^{\text{wt}(\eta)}$. As $\nu^1 \subset \lambda(\rho)$, it follows from clause (iib) of Definition 3.1.7 that $x \in A$ implies $x \in A^{\text{wt}(\eta)}$, a final contradiction proving (i).

(ii): (The idea here is that only one node of T^1, $\text{up}(\eta)$, will act to place x into a set, and once a derivative of $\text{up}(\eta)$ acts to place x into the set, that derivative is validated and remains the only derivative of $\text{up}(\eta)$ for which action can be supported at subsequent nodes.) Suppose that $\langle \eta_i, \rho_i \rangle$ share the set axiom $x \in A$ for $i \leq 1$, that $\rho_0 \subseteq \rho_1$, and that the action for $\langle \eta_0, \rho_0 \rangle$ is positive. We assume that the action for $\langle \eta_1, \rho_1 \rangle$ is negative, and obtain a contradiction. For $i \leq 1$, let $\eta_i^1 = \text{up}(\eta_i)$, and fix $\nu_i \subseteq \rho_i$ such that $\nu_i^- = \eta_i$. The construction supports consistent action, so by clauses (iic) and (iiic) of Definition 3.1.7, $\nu_1 \subset \nu_0$. Now $\eta_0^1 = \text{up}(\eta_0)$ must have an initial derivative $\mu_0 \subset \nu_1$, else by (2.8.4), Lemma 8.1.1(i) (Limit Path), and clause (iia) of Definition 3.1.7, $x \geq \text{wt}(\eta_0^1) \geq \text{wt}(\lambda(\nu_1))$, contrary to clause (iiib) of Definition 3.1.7. By clause (iid) of Definition 3.1.7, η_0 is validated, so has Π outcome, along $\rho_0 \subseteq \rho_1$, so $[\mu_0, \eta_0]$ is a primary ρ_1-link restraining $\eta_1 = \nu_1^-$. Thus η_1 is not ρ_1-free, so action for η_1 cannot be supported at ρ_1, contrary to our assumption.

(iii): (The idea here is that incompatible shared functional action can only be generated by incomparable nodes, and the node switching the paths takes action to destroy the original computation, thereby allowing a new axiom to consistently be declared.) Suppose that the hypothesis of clause (iv) of Definition 3.1.7 holds, that $\rho_0 \subseteq \rho_1$, and that $\langle \eta_0, \rho_0 \rangle$ and $\langle \eta_1, \rho_1 \rangle$ share a functional axiom with oracle A and different values. For $i \leq 1$, fix $\nu_i \subseteq \rho_i$ at which the respective functional axioms are newly defined, and note that by clause (ib) of Definition 3.1.7, $\nu_i^- = \eta_i$. Then $\langle \eta_0, \nu_0 \rangle$ and $\langle \eta_1, \nu_1 \rangle$ share this functional axiom. Hence by clause (iva) of Definition 3.1.7, $\lambda(\nu_0)|\lambda(\nu_1)$, so $\nu_0 \neq \nu_1$. Assume that $\nu_0 \subset \nu_1$. (We later show that this implies that $\rho_0 \subset \rho_1$, so by symmetry, it will not be possible to have $\nu_1 \subset \nu_0$.)

By clauses (iva,b) of Definition 3.1.7, there are derivatives $\kappa_0 \subset \kappa_1$ of κ^1 such that κ_0 has Σ outcome along ν_0 and $\kappa_1 \supseteq \nu_0$ has Π outcome (so is validated) along ν_1. Furthermore, by clause (id) of Definition 3.1.7, for $i \leq 1$, η_i is the principal derivative of η_i^1 along ρ_i, so by Lemma 8.1.1(i) (Limit Path), $\lambda(\nu_i) \subseteq \lambda(\rho_i)$. Hence $\lambda(\rho_0)|\lambda(\rho_1)$, and so $\rho_0 \subseteq \kappa_1 \subset \nu_1 \subseteq \rho_1$. (It is here that we conclude, by symmetry, that we must have $\nu_0 \subset \nu_1$.) By hypothesis and clauses (ivc,d) of Definition 3.1.7, $\mathrm{wt}(\kappa^1)$ is placed into $A^{\mathrm{wt}(\nu_1)}$. But $\kappa^1 \subset \lambda(\nu_0)$, so by (2.8.1), $\mathrm{wt}(\kappa^1) < \mathrm{wt}(\lambda(\nu_0))$; the latter is less than the use of the axiom declared for η_0 at ρ_0. ⊣

The following lemma was used in constructing a properly d-c.e. degree. It was proved as Lemma 3.4.2.

LEMMA 8.8.2 (Delayed Set Action Lemma). *Suppose that for a given faithfully executed construction, action for η is supported at ρ only when η is ρ-free, sentence decomposition is monotonic, the construction admits consistent action, and clauses (i) and (ii) of Definition* 3.4.1 *are satisfied. Then shared set action is compatible.*

PROOF. The proof is virtually the same as that of Lemma 8.8.1(ii) (Restraint). We have just stretched the action out along an interval of T^1, and no action within that interval is relevant to the set action in the compressed version. We leave the proof of this lemma to the reader. ⊣

Another version of the Restraint Lemma, under weakened hypotheses, is needed in the proof of the Sacks Splitting Theorem. We restate it here. The proof is virtually the same as that of Lemma 8.8.1(ii) (Restraint), and we leave it to the reader.

LEMMA 8.8.3 (Splitting Restraint Lemma). *Suppose that we are given a Δ_2 construction satisfying clauses (i), (iib,c,d) and (iii) of Definition* 3.1.7, *as well as the following condition:*

(*i*) *If $\eta^{2\frown}\langle\gamma^1\rangle = \nu^2 \subseteq \sigma^2 \in T^2$, η^2 has Σ outcome along ν^2, action for η^2 along $\rho^2 \supseteq \nu^2$ specifies a restraint condition for the set A, and action for σ^2 along $\tau^2 \supseteq \nu^2$ specifies that a number x be placed in the same set A, then $x \geq \mathrm{wt}(\gamma^1)$.*

Suppose also that the construction is faithfully executed, action for η is supported at ρ only when η is ρ-free, sentence decomposition is monotonic, and the construction admits consistent action. Then conclusions (i) and (ii) of Lemma 3.1.9 *(Restraint) hold. Furthermore, if hypothesis (iii) of Lemma* 8.8.1 *holds for the functional Δ, then shared functional action for Δ is compatible.* ⊣

The following Lemma was proved as Lemma 5.1.1

LEMMA 8.8.4 (Π_2-Restraint). *Suppose that for a given faithfully executed construction, action for η is supported at ρ only when η is ρ-free, sentence decomposition is monotonic, the construction admits consistent action, and clauses (i)–(iii) of Definition* 3.1.7, *with equality in (ia), are satisfied. Assume also that the following condition is satisfied for the functional Δ with oracle A.*

(*i*) *Suppose that $\langle\eta_0^2, \rho_0^2\rangle$ and $\langle\eta_0^2, \rho_0^2\rangle$ share the functional axiom $\Delta(A; x)$, and let $\kappa^2 = \eta_0^2 \wedge \eta_1^2$. Then:*
(*a*) *If $\eta_0^2 \neq \eta_1^2$, then $\eta_0^2 | \eta_1^2$.*

(*b*) *If* $\text{lev}(\kappa^2) \geq 2$, *and* $\kappa^1 \in T^1$ *is a derivative of* κ^2, *then validated action for* κ^1 *requires that* $\text{wt}(\kappa^1)$ *be placed into* A, *and no other requirement on* T^1 *has action placing* $\text{wt}(\kappa^1) \in A$.

(*c*) *For* $i \leq 1$, *if action for* η_i^1 *is supported at* ρ_i^1 *and* $\text{up}(\eta_i^1) = \eta_i^2$, *then* $\text{wt}(\eta_i^1) \leq x < \text{wt}(\rho_i^1)$.

Then shared functional action for $\Delta(A)$ *is compatible.*

PROOF. We proceed by cases, comparing the locations of η_0^j and η_1^j for $j = 1, 2$.

Case 1: $\eta_0^1 | \eta_1^1$. Let $\tau^1 = \eta_0^1 \wedge \eta_1^1$. Without lss of generality, we may assume that τ^1 has Π outcome along η_0^1 and Σ outcome along η_1^1; fix $\tau \subset \Lambda^0$ such that $\text{up}(\tau) = \tau^1$ and τ has Π outcome along Λ^0, let v be the immediate successor of τ along Λ^0, and let $v^1 = \lambda(v)$. By Lemma 8.1.1(ii) (Limit Path), $(v^1)^- = \tau^1$. By Remark 2.4.2(ii), (2.8.1) and (2.8.4), if action for η_i^1 is supported at ρ_i^1 for $i \leq 1$, then $\text{wt}(\rho_0^1) < \text{wt}(v^1) \leq \text{wt}(\eta_1^1)$. But by (ic), $\text{wt}(\eta_1^1) \leq x < \text{wt}(\rho_0^1)$, yielding a contradiction. So this case cannot occur.

Case 2: Case 1 does not apply and $\eta_0^2 = \eta_1^2$. Then without loss of generality we may assume that $\eta_0^1 \subset \eta_1^1$. By (2.6.2), η_0^1 must have Σ outcome along η_1^1. Let the shared axioms be declared at ρ_0^1 and ρ_1^1 respectively, with uses u_0 and u_1 respectively. There are several subcases.

Subcase 2.1: η_0^1 has Π outcome along ρ_0^1. Then η_0^1 must be switched at some $\xi \in [\rho_0^1, \eta_1^1]$, and as axioms are declared only at level 2 nodes, it follows from (ib) that $\text{wt}(\eta_0^1)$ is placed into A as action for η_0^1 at its immediate successor v_0^1 along η_1^1. By Lemma 8.1.1(ii) (Limit Path), (2.8.1), and condition (ia) of Definition 3.1.7 with equality instead of $\leq$, $\text{wt}(\eta_0^1) \leq u_0$, so the shared axioms are declared with incompatible oracles.

Subcase 2.2: η_0^1 and η_1^1 have Σ outcome along ρ_0^1 and ρ_1^1 respectively. In this situation, the shared axioms are compatible.

Subcase 2.3: η_0^1 has Σ outcome along ρ_0^1 and η_1^1 has Π outcome along ρ_1^1. By clause (ib) of Definition 3.1.7, $\langle \eta_0^1, v_0^1 \rangle$ also share this action, so by (ic), $x < \text{wt}(v_0^1)$. By Lemma 8.1.1(ii) (Limit Path), $v_0^1 \subseteq \eta_1^1$, so by (2.8.1) and (ic), $\text{wt}(v_0^1) \leq \text{wt}(\eta_1^1) \leq x$, a contradiction, so this situation cannot occur.

Case 3: Cases 1 and 2 do not apply. Then $\eta_0^2 \neq \eta_1^2$, so by (ia), $\eta_0^2 | \eta_1^2$. There are two subcases.

Subcase 3.1: $\text{lev}(\kappa^2) < 2$. By (2.6.3), κ^2 has a unique derivative $\kappa^1 \subset \eta_0^1, \eta_1^1$, and as Case 1 does not apply, κ^1 must have the same outcome along both η_0^1 and η_1^1. But then κ^2 has the same outcome along both η_0^2 and η_1^2, contrary to the case assumption that implies that $\eta_0^2 | \eta_1^2$.

Subcase 3.2: Otherwise. Then $\text{lev}(\kappa^2) \geq 2$. Without loss of generality, we may assume that $\eta_0^1 \subset \eta_1^1$, so κ^2 must have an initial derivative $\mu^1 \subset \eta_0^1$ that has Σ outcome along η_0^1, and a principal derivative $\pi^1 \in (\eta_0^1, \eta_1^1)$ such that π^1 has Π outcome along η_1^1. Let the shared axioms be declared at ρ_0^1 and ρ_1^1 respectively, with uses u_0 and u_1 respectively.

We first assume that $\text{out}(\rho_0^1) \subseteq \text{out}(\rho_1^1)$ and obtain a contradiction. Let π be the initial derivative of π^1 along Λ^0. Then $\text{out}(\rho_0^1) \subset \pi$, else η_0^1 would be ρ_0^1-restrained, so action for η_0^1 would not be supported at ρ_0^1. Hence by (ic), (2.8.4) and Lemma 8.1.2(i) (Out) and (2.8.1), $x < \text{wt}(\rho_0^1) \leq \text{wt}(\pi^1) \leq \text{wt}(\eta_1^1) \leq x$, a contradiction.

We thus conclude that $\text{out}(\rho_1^1) \subset \text{out}(\rho_0^1)$, and hence there must be a $\sigma \in [\text{out}(\rho_1^1), \text{out}(\rho_0^1))$ that switches π^1. As $\text{lev}(\pi^1) \geq 2$, it follows from (ib) and condition (iic) of Definition 3.1.7 that $\text{wt}(\pi^1)$ is placed into A as action at the immediate successor of σ along Λ^0. Fix $\nu_1^1 \subseteq \rho_1^1$ such that $(\nu_1^1)^- = \eta_1^1$. Hence by the equality for conditions (ia) of Definition 3.1.7, Lemma 8.1.1(ii) (Limit Path) and (2.8.1), $u_1 = \text{wt}(\nu_1^1) > \text{wt}(\pi_1^1)$, so the oracles for the shared axioms are incompatible. We conclude that the shared functional action is compatible. ⊣

There is an additional restraint lemma that was used at the Δ_3 level. As it involves backtracking, it will be proved in the next chapter.

8.9. Miscellaneous Lemmas

We recall the following definitions from Section 5.1

DEFINITION 8.9.1. Fix $\eta^2 \in T^2$ such that $|\eta^2| = 2e+2$. Define the string χ_{η^2} of length $e+1$ by $\chi_{\eta^2}(x) = 0$ ($= 1$, resp.) iff $\eta^2 \upharpoonright 2x$ is activated (validated, resp.) along η^2.

DEFINITION 8.9.2. Fix $\nu^1 \in T^1$. Define the string γ_{ν^1} of length $|\nu^1|$ by $\gamma_{\nu^1}(x) = 0$ ($= 1$, resp.) iff $\nu^1 \upharpoonright x$ is activated (validated, resp.) along ν^1.

The following lemma was introduced as Lemma 5.1.5.

LEMMA 8.9.3 (Incompatible Oracle Lemma). *Let $\xi \subset \eta \subset \Lambda^0 \in [T^0]$ and $\eta^2 \neq \xi^2 \in T^2$ be given such that $\xi = \text{out}^0(\xi^2)$, $\eta = \text{out}^0(\eta^2)$, and $|\eta^2| = |\xi^2| = 2e+2$. Suppose that for all $\rho^2 \in T^2$ such that $|\rho^2|$ is odd, $\text{lev}(\rho^2) < 2$. Then either χ_{η^2} and χ_{ξ^2} are incompatible, or $\gamma_{\text{out}(\eta^2)}$ and $\gamma_{\text{out}(\xi^2)}$ are incompatible.*

PROOF. *Case* 1: $\eta^1 = \text{out}(\eta^2)$ and $\xi^1 = \text{out}(\xi^2)$ are compatible. By Lemma 8.1.2(i) (Out), $\eta^1 \neq \xi^1$, else $\eta^2 = \lambda(\eta^1) = \lambda(\xi^1) = \xi^2$. As $\xi \subset \eta$, it can be shown that our assumption implies that $\xi^1 \subset \eta^1$; however, we will assume this to be the case without loss of generality, as a symmetrical proof will hold if $\eta^1 \subset \xi^1$. Let $\rho^2 = \eta^2 \wedge \xi^2$. As $\eta^2 \neq \xi^2$ and $|\eta^2| = |\xi^2|$, $\rho^2 \subset \eta^2, \xi^2$. Fix $i = 1$. (We do this for the sake of the next paragraph, as that paragraph will also apply when $i = 0$ for the next case.)

By Remark 2.4.2(vi), ρ^{i+1} has an initial derivative $\mu^i \subset \xi^i$ and a principal derivative $\pi^i \subset \eta^i$ such that $\xi^i \subset \pi^i$, μ^i has Σ outcome along ξ^i, and π^i has Π outcome along η^i. Thus by Remark 2.4.2(v), ρ^{i+1} has Π outcome along ξ^{i+1} and Σ outcome along η^{i+1}.

Now as $\mu^1 \neq \pi^1$, it follows from (2.6.3) that $\text{lev}(\rho^2) = 2$, and so that $|\rho^2|$ is even. Hence χ_{η^2} and χ_{ξ^2} are incompatible.

Case 2: $\eta^1 = \text{out}(\eta^2)$ and $\xi^1 = \text{out}(\xi^2)$ are incompatible. Let $\rho^1 = \eta^1 \wedge \xi^1$, and note that by the case assumption, $\rho^1 \subset \eta^1, \xi^1$. We now set $i = 0$ for the second paragraph of Case 1, and conclude by that paragraph that ρ^1 has Π outcome along ξ^1 and has Σ outcome along η^1. Thus $\gamma_{\text{out}(\eta^2)}$ and $\gamma_{\text{out}(\xi^2)}$ are incompatible. ⊣

Chapter 9

BACKTRACKING

The concept of backtracking was introduced in [13]. Backtracking is a procedure that takes a non-$\lambda^k(\xi)$-free node of T^k, and returns it to $\lambda^k(\eta)$-free status for some $\eta \supset \xi$, with the same initial derivative. The concept, as manifested in [13], was intertwined with the validity of certain sentences, and so its analysis was substantially more complicated than we need, as we look at it without regard to the requirements assigned to the nodes. The lemmas we prove here can also be generally applied to systems of trees, as there is no dependence on the requirements to be satisfied.

This chapter is devoted to proving lemmas describing the behavior of the backtracking process. Throughout the chapter, we assume that we are dealing with a level ℓ construction.

9.1. Backtracking

Backtracking cannot always be carried out. In this section, we identify the sole obstruction to backtracking, and define a sequence generated by primary links that determines how backtracking is to be implemented, when possible.

Definition 9.1.1. Fix $m \leq \ell, \sigma, \xi \in T^0$, and $\sigma^m \in T^m$. We say that $\langle \sigma^m, \sigma \rangle$ is *weakly ξ-accessible* if $\text{init}(\sigma^m, \sigma) = \sigma$ and there is an $\eta \in T^0$ such that $\eta \supseteq \xi$, $\text{init}(\sigma^m, \eta) = \sigma$ and $\sigma^m \subseteq \lambda^m(\eta)$; and if this is the case, we call η a *witness to the weak ξ-accessibility of* $\langle \sigma^m, \sigma \rangle$. We say that $\langle \sigma^m, \sigma \rangle$ is *ξ-accessible* if there is an $\eta \in T^0$ such that η is a witness to the weak ξ-accessibility of $\langle \sigma^m, \sigma \rangle$ and $\text{up}^m(\eta) = \sigma^m$.

We will prove that $\langle \sigma^m, \sigma \rangle$ is weakly ξ-accessible, then iff there is no primary ξ-link restraining σ. The next lemma proves one direction.

Lemma 9.1.2 (Nonaccessibility Lemma). *Fix $m \leq \ell, \xi \subset \eta \in T^0$ and $\sigma^m \in T^m$ such that $\sigma^m \subset \lambda^m(\xi)$, and for all $i \leq m$, let $\sigma^i = \text{init}^i(\sigma^m, \xi)$ and set*

$\sigma = \sigma^0$. *Suppose that there is a primary* η*-link* $[\mu, \pi]$ *that properly restrains* σ. *Then* $\langle \sigma^m, \sigma \rangle$ *is not weakly* η*-accessible.*

PROOF. By Lemma 8.3.3(iii) (Link Analysis), $\sigma^1 \not\subseteq \lambda(\rho)$ for any $\rho \supseteq \eta$. Hence by Lemma 8.1.1(i) (Limit Path), if $\rho \supseteq \eta$ and $\sigma^m \subset \lambda^m(\rho)$, then σ^1 cannot be the initial derivative of σ^m along $\lambda(\rho)$, so σ cannot be the initial derivative of σ^m along ρ. ⊣

We now begin an analysis of the links that must be switched in order to return σ^m to the current path through T^m as a free node, beginning at $\eta \in T^0$, with specified initial derivative σ, when there is no primary η-link that properly restrains σ. It will be possible to return σ^m to free status with initial derivative σ iff $\langle \sigma^m, \sigma \rangle$ is weakly η-accessible. In this case, we will show that there is a set of links that we can switch to return σ^m to free status with initial derivative σ, and that this set of links must always be switched if we are to return σ^m to free status with initial derivative σ; and we will present an algorithm to return σ^m to free status with initial derivative σ. We begin by defining the set of links that must be switched in order to do so.

DEFINITION 9.1.3. Fix $m \leq \ell$, $\sigma^m \in T^m$ and $\xi \subseteq \eta \in T^0$ such that σ^m is $\lambda^m(\xi)$-free. We define a sequence $\mathrm{PL}(\sigma^m, \xi, \eta)$ and strings $\mu(\sigma^m, \xi, \eta)$ and $\pi(\sigma^m, \xi, \eta)$ by induction on $\eta \supseteq \xi$. We begin by setting $\mathrm{PL}(\sigma^m, \xi, \xi) = \langle \sigma^m \rangle$ and $\mu(\sigma^m, \xi, \xi) = \pi(\sigma^m, \xi, \xi) = \sigma^m$. Suppose that $\eta \supset \xi$ and that $\mathrm{PL}(\sigma^m, \xi, \eta^-)$, $\mu(\sigma^m, \xi, \eta^-)$ and $\pi(\sigma^m, \xi, \eta^-)$ have been defined, and that $\pi(\sigma^m, \xi, \eta^-) \in T^r$. Define $\pi^j(\sigma^m, \xi, \eta^-)$ to be the principal derivative of $\pi(\sigma^m, \xi, \eta^-)$ along $\lambda^j(\eta^-)$ and $\mu^j(\sigma^m, \xi, \eta^-)$ to be the initial derivative of $\mu(\sigma^m, \xi, \eta^-)$ along $\lambda^j(\eta^-)$ if $j < r$ whenever these strings exist; and let $\mu^j(\sigma^m, \xi, \eta^-) = \pi^j(\sigma^m, \xi, \eta^-) = \mathrm{up}^j(\pi^r(\sigma^m, \xi, \eta^-))$ for $j \in [r, \ell]$. There are three cases, and we follow the first case that applies:

Case 1: η switches $\pi(\sigma^m, \xi, \eta^-)$. If $|\mathrm{PL}(\sigma^m, \xi, \eta^-)| = 1$, set $\mathrm{PL}(\sigma^m, \xi, \eta) = \langle \sigma^m \rangle$ and $\pi(\sigma^m, \xi, \eta) = \mu(\sigma^m, \xi, \eta) = \sigma^m$. Otherwise, $|\mathrm{PL}(\sigma^m, \xi, \eta^-)| > 1$ and we set $\mathrm{PL}(\sigma^m, \xi, \eta) = (\mathrm{PL}(\sigma^m, \xi, \eta^-))^-$. Fix the longest $\nu \subset \eta$ such that $\mathrm{PL}(\sigma^m, \xi, \eta) = \mathrm{PL}(\sigma^m, \xi, \nu)$ and set $\pi(\sigma^m, \xi, \eta) = \pi(\sigma^m, \xi, \nu)$ and $\mu(\sigma^m, \xi, \eta) = \mu(\sigma^m, \xi, \nu)$.

Case 2: Case 1 does not apply and there are $s \leq \ell$ and $\mu^s, \pi^s \in T^s$ such that $[\mu^s, \pi^s]$ is a primary $\lambda^s(\eta)$-link and for some $j \leq s$, the $\lambda^j(\eta)$-link $[\mu^j, \pi^j]$ derived from $[\mu^s, \pi^s]$ restrains $\pi^j(\sigma^m, \xi, \eta^-)$. Fix the smallest such s and the longest such link for s (its existence follows from Lemma 8.3.1 (Nesting)), define $\mathrm{PL}(\sigma^m, \xi, \eta) = \mathrm{PL}(\sigma^m, \xi, \eta^-)^\frown \langle \pi^s \rangle$, $\mu(\sigma^m, \xi, \eta) = \mu^s$ and $\pi(\sigma^m, \xi, \eta) = \pi^s$.

Case 3: Neither Case 1 nor Case 2 applies. Then we set $\mathrm{PL}(\sigma^m, \xi, \eta) = \mathrm{PL}(\sigma^m, \xi, \eta^-)$, $\pi(\sigma^m, \xi, \eta) = \pi(\sigma^m, \xi, \eta^-)$ and $\mu(\sigma^m, \xi, \eta) = \mu(\sigma^m, \xi, \eta^-)$.

9.2. Properties of PL Sets

The next lemma is the key to understanding the evolution of the PL sequences. It implies that the sequences evolve in an organized way. For example, the last node is always free. Also, if the sequence computed at ξ is identical to that computed at $\eta \supset \xi$, then the outcome of the last node of the sequence along the appropriate path computed from η is the same as the outcome along the similar path computed from ξ, and the initial derivative of the last node of the sequence along η is the initial derivative of that node along ξ. Furthermore, the intervals formed from the strings in the sequence are successively nested.

LEMMA 9.2.1 (PL Analysis Lemma). *Fix $m \le \ell$, $\sigma^m \in T^m$ and $\xi \subseteq \eta \in T^0$ such that σ^m is $\lambda^m(\xi)$-free, and let $\sigma = \mathrm{init}(\sigma^m, \xi)$. Let $\mathrm{PL}(\sigma^m, \xi, \eta) = \langle \pi_i^{k_i} : i \le n\rangle$ and $\mathrm{PL}(\sigma^m, \xi, \eta^-) = \langle \pi_i^{k_i} : i \le \widetilde{n}\rangle$ (one sequence is an end extension of the other) where $\pi_i^{k_i} \in T^{k_i}$, and let $\widehat{n} = \max\{n, \widetilde{n}\}$. For all $i \le \widehat{n}$, fix the shortest $\theta_i \in [\xi, \eta]$ such that $|\mathrm{PL}(\sigma^m, \xi, \rho)| \ge i+1$ for all $\rho \in [\theta_i, \eta)$, and fix the longest $\widetilde{\theta}_i \in [\xi, \eta)$ such that $|\mathrm{PL}(\sigma^m, \xi, \widetilde{\theta}_i)| = i+1$. For all $i \le \widehat{n}$, let $\mu_i^{k_i} = \mu(\sigma^m, \xi, \theta_i)$ and $\widehat{\mu}_i = \mathrm{init}(\pi_i^{k_i}, \theta_i)$, for all $j \le k_i$ let $[\mu_i^j, \pi_i^j]$ be the $\lambda^j(\theta_i)$-link derived from $[\mu_i^{k_i}, \pi_i^{k_i}]$ (if $j = 0$, we omit the superscript), and for all $j \in (k_i, \ell]$ let $\pi_i^j = \mathrm{up}^i(\pi_i^{k_i})$. For all $j \le \ell$, let $\kappa^j = \lambda^j(\eta^-) \wedge \lambda^j(\eta)$. Then:*

(i) *For all $j \le \ell$, π_n^j is $\lambda^j(\eta)$-free.*
(ii) *If $0 < n \le \widetilde{n}$, then $\lambda^j(\widetilde{\theta}_n) \wedge \lambda^j(\eta) \supset \pi_n^j$ for all $j \le \ell$.*
(iii) *If $n = 0$, then $\mathrm{init}(\sigma^m, \eta) = \sigma$.*
(iv) *If $\widetilde{n} < n$, then $\pi_n^{k_n} = \kappa^{k_n}$.*
(v) *For all $i < n$ and $j \le \min\{k_{i+1}, k_i\}$, $\mu_{i+1}^j \subseteq \mu_i^j \subseteq \pi_i^j \subset \pi_{i+1}^j$. Furthermore, if $k_{i+1} \le k_i$, then $\mu_{i+1}^j \subset \mu_i^j$.*
(vi) *Suppose that $[\overline{\mu}^r, \overline{\pi}^r]$ is a primary $\lambda^r(\eta)$-link whose derived η-link $[\overline{\mu}, \overline{\pi}]$ η-restrains σ. Then there is a $q \le r$, a primary $\lambda^q(\eta)$-link $[\widetilde{\mu}^q, \widetilde{\pi}^q]$ with derived $\lambda^i(\eta)$-links $[\widetilde{\mu}^i, \widetilde{\pi}^i]$ for $i \le q$, an $i \in [1, n]$ and a derivative $\widetilde{\pi}^{k_i} \in T^{k_i}$ of $\widetilde{\pi}^r$ such that $\widetilde{\pi}^q$ is $\lambda^q(\theta_i)$-free, $\widetilde{\mu}^0 \subseteq \overline{\mu}^0 \subset \overline{\pi}^0 \subseteq \widetilde{\pi}^0$, $\pi_i^{k_i} = \widetilde{\pi}^{k_i}$ and $\mathrm{init}(\widetilde{\pi}^{k_i}, \eta)\downarrow = \widehat{\mu}_i$. Furthermore, if $\overline{\pi}^r$ is $\lambda^r(\eta)$-free, then we can assume that $[\widetilde{\mu}^q, \widetilde{\pi}^q] = [\overline{\mu}^r, \overline{\pi}^r]$.*

PROOF. We proceed by induction on $|\eta| - |\xi|$. If $\eta = \xi$, then by hypothesis, $\sigma^m = \pi_0^m$ is $\lambda^m(\xi)$-free, so by Lemma 8.5.1(i,ii) (Freeness Transmission), π_0^j is $\lambda^j(\eta)$-free for all $j \le \ell$. Hence (i) holds. (ii)–(vi) are immediate.

Suppose that $\eta \supset \xi$. We follow the case structure of Definition 9.1.3 to prove (i)–(v), noting that it is easy to see that for all $i \le \widetilde{n}$,

$$\mathrm{PL}(\sigma^m, \xi, \eta^-) = (\mathrm{PL}(\sigma^m, \xi, \widetilde{\theta}_i))^{-\frown} \mathrm{PL}(\pi_i^{k_i}, \widetilde{\theta}_i, \eta^-). \qquad (*)$$

(vi) is proved at the end. Let $k = k_n$.

Case 1: η switches $\pi_{\widetilde{n}}^{k_{\widetilde{n}}}$. If $\widetilde{n} = 0$, then (i) follows from Lemma 8.5.1(iii) (Freeness Transmission), (iii) follows by definition, (i) and induction, and (ii), (iv) and (v) are vacuous.

Otherwise, $\widetilde{n} > 0$. ($\mathrm{PL}(\sigma^m, \xi, \eta) = (\mathrm{PL}(\sigma^m, \xi, \eta^-))^-$ in this case.) Let $\nu = \theta_{\widetilde{n}}$. Note that $\pi(\sigma^m, \xi, \nu) = \pi(\sigma^m, \xi, \eta^-)$. For $j \leq \ell$ and $\rho \in \{\nu, \eta\}$, let $\kappa_\rho^j = \lambda^j(\rho^-) \wedge \lambda^j(\rho)$. We note that by Lemma 8.5.1(iii) (Freeness Transmission), for $j \leq \ell$ and $\rho \in \{\nu, \eta\}$, κ_ρ^j is both $\lambda^j(\rho^-)$-free and $\lambda^j(\rho)$-free. For all $j \leq \ell$ and $\rho \in \{\nu^-, \nu, \eta^-, \eta\}$, let $\pi_\rho^j = \pi^j(\sigma^m, \xi, \rho)$. Let $r = k_{\widetilde{n}}$ and for all $j \leq r$, let $\mu^j = \mu_{\widetilde{n}}^j$ and $\pi^j = \pi_{\widetilde{n}}^j = \pi_\nu^j$.

Before verifying (i)–(v), we show that $\pi_{\nu^-}^j$ is $\lambda^j(\eta)$-free for all $j \leq \ell$. For assume that this is not the case, in order to obtain a contradiction. Then by Lemma 8.3.1 (Nesting), there is a $j \leq \ell$ such that $\pi_{\nu^-}^j$ is restrained by a $\lambda^j(\eta)$-link derived from a primary $\lambda^t(\eta)$-link $[\widetilde{\mu}^t, \widetilde{\pi}^t]$ for which $\widetilde{\pi}^t$ is $\lambda^k(\eta)$-free. Fix such a link with t minimal. For all $j \leq t$, let $[\widetilde{\mu}^j, \widetilde{\pi}^j]$ be the $\lambda^j(\eta)$-link derived from $[\widetilde{\mu}^t, \widetilde{\pi}^t]$ (we omit the superscript if $j = 0$). We proceed by subcases.

Subcase 1.1: $t \geq r$ and $\widetilde{\pi}^s = \kappa_\eta^s$ for some s. As $\mathrm{up}^t(\kappa_\eta^s) = \kappa_\eta^t$ and $\mathrm{up}^t(\widetilde{\pi}^s) = \widetilde{\pi}^t$, $\widetilde{\pi}^t = \kappa_\eta^t$. But $[\mu_{\eta^-}^r, \kappa_\eta^r]$ is a primary $\lambda^r(\eta^-)$-link, so Lemma 8.5.8(ii) (Alternating Link) and the minimality of t rule out this possibility.

Subcase 1.2: Subcase 1.1 does not apply, and $\widetilde{\pi}^s \subset \kappa_\eta^s$ for some $s \geq r$. As $s \geq r$, $\kappa_\nu^s = \kappa_\eta^s$, and we have noted that κ_η^s is both $\lambda^s(\eta)$-free and $\lambda^s(\nu^-)$-free. Furthermore, by Lemma 8.5.1(ii) (Freeness Transmission), $\widetilde{\pi}^s$ is $\lambda^s(\eta)$-free. By (ii) inductively, κ_η^s has Σ outcome along $\lambda^s(\nu)$ iff κ_η^s has Σ outcome along $\lambda^s(\eta^-)$, so as both η and ν switch $\kappa_\nu^s = \kappa_\eta^s$, κ_η^s has Σ outcome along $\lambda^s(\nu^-)$ iff κ_η^s has Σ outcome along $\lambda^s(\eta)$. By Lemma 8.5.7 (Stable Link), $[\widetilde{\mu}^j, \widetilde{\pi}^j]$ is also a $\lambda^j(\nu^-)$-link for all $j \in [s, t]$. Now as $\widetilde{\pi}^s \subset \kappa_\eta^s$, $\widetilde{\pi}^s$ has the same outcome, δ^s, along both $\lambda^s(\nu^-)$ and $\lambda^s(\eta)$, so $(\mathrm{out}^j(\delta^s))^- = \widetilde{\pi}^j$ is the principal derivative of $\widetilde{\pi}^s$ along both $\lambda^j(\eta)$ and $\lambda^j(\nu^-)$ for all $j \leq s$. Thus $[\widetilde{\mu}^j, \widetilde{\pi}^j]$ is a $\lambda^j(\nu^-)$-link for all $j \leq t$, so $\pi_{\nu^-}^j$ is not $\lambda^j(\nu^-)$-free for some $j \leq t$, contradicting (i) inductively.

Subcase 1.3: $t < r$ and $\widetilde{\pi}^t \subset \kappa_\eta^t$. Then $[\widetilde{\mu}^t, \widetilde{\pi}^t]$ is also a $\lambda^t(\eta^-)$-link and i-envelops $\pi_{\nu^-}^k$ for some i. Note that by (*), $\mathrm{PL}(\pi_{\nu^-}^k, \nu^-, \eta^-) = \langle \pi_{\nu^-}^k, \pi_\nu^r \rangle$, so by (vi) inductively, $r \leq t$, a contradiction.

Subcase 1.4: The previous subcases do not apply and there is an $i < r$ such that $\widetilde{\mu}^{i+1} \supset \kappa_\eta^{i+1}$ but $\widetilde{\pi}^j \subset \kappa_\eta^j$ for all $j \leq i$. Then by Lemma 8.5.3 (Crossover) and Lemma 8.5.1(iii) (Freeness Transmission), there is a primary $\lambda^i(\eta^-)$-link $[\widehat{\mu}^i, \kappa_\eta^i]$ such that $\widehat{\mu}^i \subset \widetilde{\mu}^i \subset \widetilde{\pi}^i \subset \kappa_\eta^i$ and κ_η^i is $\lambda^i(\eta^-)$-free, so by Lemma 8.5.2 (Link Comparison), $[\widehat{\mu}^i, \kappa_\eta^i]$ j-envelops $\pi_{\nu^-}^k$ along $\lambda^j(\eta^-)$ for some $j \leq i$. Note that by (*), $\mathrm{PL}(\pi_{\nu^-}^k, \nu^-, \eta^-) = \langle \pi_{\nu^-}^k, \pi_\nu^r \rangle$, so by (vi) inductively, $r \leq i$, a contradiction.

Subcase 1.5: The preceding cases do not apply and $t > r$. As κ_η^t is $\lambda^t(\eta)$-free, $\kappa_\eta^t \subset \widetilde{\mu}^t$. As $\kappa_\eta^t = \kappa_\nu^t$ and the preceding cases do not apply, it follows from

Lemma 8.5.3 (Crossover) and as $\widetilde{\pi} \subset \eta^- = \kappa^0_\eta$ that there is an $i \in [r, t-1]$ such that $\widetilde{\pi}^i \subset \kappa^i_\eta$. Thus Subcase 1.2 would be followed.

Subcase 1.6: Otherwise. Then $t < r$ and $\widetilde{\pi}^s = \kappa^s$ for some $s \leq t$. As $\kappa^t = \text{up}^t(\kappa^s)$ and $\widetilde{\pi}^t = \text{up}^t(\widetilde{\pi}^s)$, $\widetilde{\pi}^t = \kappa^t$. By Lemma 8.5.8(i,ii) (Alternating Link), $s < r$ and there is a $j < t$ and an interval $[\widehat{\mu}^j, \widehat{\pi}^j]$ that is both a primary $\lambda^j(\eta^-)$-link and a primary $\lambda^j(\eta)$-link for which $\widehat{\pi}^j$ is $\lambda^r(\eta^-)$-free. Note that by (*), $\text{PL}(\pi^k_{\nu^-}, \nu^-, \eta^-) = \langle \pi^k_{\nu^-}, \pi^r_\nu \rangle$, so by (vi) inductively, $r \leq j$, a contradiction.

We conclude that $\pi^k_{\nu^-}$ is $\lambda^k(\eta)$-free. Now $\pi^k_\eta = \pi^k_{\nu^-}$, so (i) follows from Lemma 8.5.1(i,ii) (Freeness Transmission). (ii) is follows immediately from (i) and Lemma 8.5.9 (Outcome Restoration), and (iii) follows easily from (ii). (iv) is vacuous and (v) follows by induction, as $\text{PL}(\sigma^m, \xi, \eta) = \text{PL}(\sigma^m, \xi, \nu^-)$.

Case 2: Case 2 of Definition 9.1.3 is followed at η. (i) and (iv) follow from Lemmas 8.5.5(iii) (Link Development) and 8.5.1(iii) (Freeness Transmission). (iii) is immediate, and (ii) is vacuous and as $n > \widehat{n}$.

We now prove (v). By Lemma 8.5.5(iii) (Link Development) and (i) inductively, $\pi^{k_n}_n = \kappa^{k_n}$ and by (i), $\pi^{k_n}_n$ is $\lambda^{k_n}(\eta)$-free. Fix $i \leq k_n$ such that $[\mu^{k_n}_n, \pi^{k_n}_n]$ i-envelops $\pi^{k_{n-1}}_{n-1}$. By Lemma 8.5.2 (Link Comparison), we must have $i \geq \min\{k_{n-1}, k_m\}$. By choice of i, $\mu^i_n \subseteq \pi^i_{n-1} \subset \pi^i_n$. Furthermore, $\mu^j_n \subseteq \pi^j_{n-1} \subset \pi^j_n$ for all $j \leq i$, else By Lemma 8.3.7 (iii) (Free and True), there is a $p < k_n$ such that a primary $\lambda^p(\eta)$-link j-envelops $\pi^{k_{n-1}}_{n-1}$ for some j; but then by Lemma 8.5.5(iii) (Link Development) and Lemma 8.5.1(iii) (Freeness Transmission), there would be a $q \leq p$ and a primary $\lambda^q(\eta)$-link whose last node is $\lambda^q(\eta)$-free that j-envelops $\pi^{k_{n-1}}_{n-1}$ for some j, contrary to the choice of k_n.

By (i), for all $j \leq k_n$, κ^j is a $\lambda^j(\eta)$-free derivative of κ^{k_n}, and by definition, π^j_n is the shortest $\lambda^j(\eta)$-free derivative of κ^{k_n}; hence $\pi^j_n \subseteq \kappa^j$. By (i) and Lemma 8.5.1(iii) (Freeness Transmission), for all $j \leq n$, both π^j_{n-1} and κ^j are $\lambda^j(\eta^-)$-free, so must be comparable; and as $\pi^j_n \subseteq \kappa^j$, π^j_n and π^j_{n-1} must be comparable for all $j \leq k_n$. Furthermore, if $j \in (k_n, \ell]$, then $\pi^j_n = \text{up}^j(\pi^{k_n}_n) = \text{up}^j(\kappa^{k_n}) = \kappa^j$, so π^j_n and π^j_{n-1} must be comparable for all $j \leq \ell$. We proceed by cases, showing that $\mu^i_n \subseteq \mu^i_{n-1}$, by comparing π^{i+1}_{n-1} and π^{i+1}_n.

First suppose that $\pi^{i+1}_{n-1} \subset \pi^{i+1}_n$. If $k_{n-1} < i$, then $\mu^i_{n-1} = \pi^i_{n-1}$ and so (v) holds. Otherwise, by choice of i and Lemma 8.5.2 (Link Comparison), $\pi^{i+1}_{n-1} \subset \mu^{i+1}_n$ and so π^i_{n-1} must be the principal derivative of π^{i+1}_{n-1} along μ^i_n, i.e., $\pi^i_{n-1} \subset \mu^i_n$, a contradiction.

Next suppose that $\pi^{i+1}_{n-1} = \pi^{i+1}_n$. As $\pi^j_n = \kappa^j$ for all $j \in [k_n, \ell]$, it follows from Lemma 8.5.8(i) (Alternating Link) that we cannot have $k_{n-1} \geq k_n$, else Case 1 of Definition 9.1.3 would apply. Hence $k_{n-1} < k_n$, so by Lemma 8.5.8(ii) (Alternating Link), there is a primary $\lambda^j(\eta)$-link $[\widehat{\mu}^j, \widehat{\pi}^j]$ for some $j < k_n$ which p-envelops $\pi^{k_n}_{n-1}$ with $\widehat{\pi}^j$ $\lambda^j(\eta)$-free, contrary to the choice of k_n. Hence this case cannot occur.

Finally, suppose that $\pi_{n-1}^{i+1} \supset \pi_n^{i+1}$, noting that $i \leq k_n$. We must have $\pi_n^{i+1} = \kappa^{i+1}$, else by Lemma 8.5.2 (Link Comparison), $\pi_n^i \subset \pi_{n-1}^i$, yielding a contradiction. Thus by Lemma 8.5.3 (Crossover), $\mu_n^i \subset \mu_{n-1}^i \subseteq \pi_{n-1}^i \subset \pi_n^i$. As $\mu_{n-1}^i \subset \pi_{n-1}^i$, it follows from by Lemma 8.5.2 (Link Comparison) that $\mu_n^j \subset \mu_{n-1}^j \subset \pi_{n-1}^j \subset \pi_n^j$ for all $j \leq i$, and so (v) holds.

Case 3: Neither Case 1 nor Case 2 applies. Then Case 3 of Definition 9.1.3 must be followed. Fix r such that $\pi(\sigma^m, \xi, \eta^-) \in T^r$. As neither Case 1 nor Case 2 applies, it follows from Lemma 8.3.3(id) (Link Analysis) that $\pi(\sigma^m, \xi, \eta^-) = \pi_{\eta^-}^r \subseteq \lambda^r(\eta)$. As $\pi_{\eta^-}^r$ is not restrained by a $\lambda^r(\eta)$-link, it must be $\lambda^r(\eta)$-free. Furthermore, $\pi(\sigma^m, \xi, \eta) = \pi_{\eta^-}^r$. (i) now follows from Lemma 8.5.1(i,ii) (Freeness Transmission).

(ii) and (iii) are immediate from the case assumption, (iv) is vacuous, and (v) follows immediately by induction.

(vi): Suppose that $[\overline{\mu}^q, \overline{\pi}^q]$ is a primary $\lambda^q(\eta)$-link whose derived η-link $[\overline{\mu}, \overline{\pi}]$ η-restrains σ. By (v), $[\mu_n, \pi_n]$ also η-restrains σ, and by (i), π_n is η-free; hence by Lemma 8.3.1 (Nesting), $[\overline{\mu}, \overline{\pi}] \subseteq [\mu_n, \pi_n]$. If $\overline{\pi} = \pi_n$, then by the minimality of the choice of k_n, we can assume without loss of generality that $[\overline{\mu}, \overline{\pi}] = [\mu_n, \pi_n]$.

Suppose that $\overline{\pi} \subset \pi_n$. We may assume that $q < k_n$, else there is nothing to show. By Lemma 8.5.2 (Link Comparison), letting $[\overline{\mu}^j, \overline{\pi}^j]$ be the $\lambda^j(\eta)$-link derived from $[\overline{\mu}^q, \overline{\pi}^q]$ for all $j \leq q$, we have $[\overline{\mu}^j, \overline{\pi}^j] \subseteq [\mu_n^j, \pi_n^j]$ and $\overline{\pi}^j \subset \pi_n^j$ for all $j \leq q$. We consider the situation at η in Definition 9.1.3, proceeding by induction on $|\eta|$.

First assume that Case 1 of Definition 9.1.3 is followed at η, and let $\nu = \theta_{\widetilde{n}}$. Then $\mathrm{PL}(\sigma^m, \xi, \nu^-) = \mathrm{PL}(\sigma^m, \xi, \eta)$, and by (i), $[\mu_n^{k_n}, \pi_n^{k_n}]$ is a primary $\lambda^{k_n}(\nu^-)$-link. As $\overline{\pi}^q \subset \pi_n^q$ and $q < k_n$, $[\overline{\mu}^q, \overline{\pi}^q]$ is a primary $\lambda^q(\nu^-)$-link, so (vi) follows by induction in this case.

Next assume that Case 2 of Definition 9.1.3 is followed at η. For $j \leq k_n$, π_n^j is the shortest $\lambda^j(\eta)$-free derivative of $\kappa_\eta^{k_n+1}$, so by Lemma 8.5.1(iii) (Freeness Transmission), $\pi_n^j \subseteq \kappa_\eta^j$ for all $j \leq k_n$. As $q < k_n$ and $\overline{\pi}^j \subset \pi_n^j$ for all $j \leq q$, $[\overline{\mu}^q, \overline{\pi}^q]$ is a primary $\lambda^q(\eta^-)$-link, so (vi) follows by induction in this case.

Finally, assume that Case 3 of Definition 9.1.3 is followed at η. By (i), $[\mu_n^{k_n}, \pi_n^{k_n}]$ is also a $\lambda^{k_n}(\eta^-)$-link. As $q < k_n$ and $\overline{\pi}^j \subset \pi_n^j$ for all $j \leq q$, $[\overline{\mu}^q, \overline{\pi}^q]$ is a primary $\lambda^q(\eta^-)$-link, so (vi) follows by induction in this case. $\dashv$

9.3. Extenders

The backtracking process is an iterative one, and entails taking extensions of $\sigma \in T^0$ of two types. We now define these extenders of σ. The definitions apply to trees of all levels.

DEFINITION 9.3.1. Fix $k < \ell$, $\eta \subset \kappa \in T^0$ and $\pi^k \in T^k$. We say that κ is a *nonswitching extender of η for π^k* if every σ such that $\eta \subset \sigma \subseteq \kappa$ is nonswitching, $\mathrm{up}^k(\kappa) = \pi^k$, $\mathrm{init}(\mathrm{up}^{k-1}(\kappa), \kappa) = \kappa$, and there is no τ such that $\eta \subset \tau \subset \kappa$, $\mathrm{up}^k(\tau) = \pi^k$, and $\mathrm{init}(\mathrm{up}^{k-1}(\tau), \tau) = \tau$. We say that κ is a *switching extender of η for π^k* if κ^- is a non-switching extender of η for π^k and κ switches π^k.

We are now ready to define backtracking extenders.

DEFINITION 9.3.2. Fix $m \leq \ell$, $\sigma^m \in T^m$ and $\xi \subseteq \eta \subset \kappa \in T^0$ such that σ^m is $\lambda^m(\xi)$-free and $\mathrm{init}(\sigma^m, \xi) \downarrow$. For $i \leq n$, fix $\pi_i^{k_i} \in T^{k_i}$ such that $\mathrm{PL}(\sigma^m, \xi, \eta) = \langle \pi_0^{k_0}, \dots, \pi_n^{k_n} \rangle$. Then we call κ a *backtracking extender of η for σ^m from ξ* if there is a sequence of nodes $\eta = \rho_{-1} \subset \rho_0 \subset \cdots \subset \rho_n = \kappa$ for which the following conditions hold:

(i) $\mathrm{up}^m(\kappa) = \sigma^m$ and $\mathrm{init}(\sigma^m, \kappa) = \mathrm{init}(\sigma^m, \xi)$.
(ii) For all $i \in [0, n-1]$, ρ_i is a switching extender of ρ_{i-1} for $\pi_{n-i}^{k_{n-i}}$.
(iii) ρ_n is a nonswitching extender of ρ_{n-1} for σ^m.

The next lemma characterizes the backtracking process.

LEMMA 9.3.3 (Backtracking Lemma). *Fix $m \leq \ell$, $\sigma^m \in T^m$ and $\xi \subseteq \eta \subset \kappa \in T^0$ such that σ^m is $\lambda^m(\xi)$-free and* $\mathrm{init}(\sigma^m, \xi) \downarrow$. *For $i \leq n$, fix $\pi_i^{k_i} \in T^{k_i}$ such that $\mathrm{PL}(\sigma^m, \xi, \eta) = \langle \pi_0^{k_0}, \dots, \pi_n^{k_n} \rangle$, and assume that $k_i > 0$ for all $i \leq n$. Let θ be the shortest string in $(\xi, \eta]$ such that $|\mathrm{PL}(\sigma^m, \xi, \theta)| > 1$ for all $\rho \in [\theta, \eta]$, and assume that σ^m has Π outcome along $\lambda^m(\theta^-)$. Then there is a backtracking extender κ of η for σ^m from ξ. Furthermore, there is a canonical way of obtaining κ.*

PROOF. We proceed by induction on $j \leq n$. We will have, as an induction hypothesis, that

$$\pi_{n-j}^{k_{n-j}} \text{ is } \lambda^{k_{n-j}}(\rho_{j-1})\text{-free.}$$

We begin by setting $\rho_{-1} = \eta$ and noting that, by Lemma 9.2.1(i) (PL Analysis) or hypothesis, $\pi_n^{k_n}$ is $\lambda^{k_n}(\eta)$-free; that by hypothesis $k_n > 0$; and that by Definition 9.1.3 and the hypothesis stating that $\sigma^m = \pi_0^{k_0}$ has Π outcome along $\lambda^m(\theta^-)$, $\pi_n^{k_n}$ has Π outcome along $\lambda^{k_n}(\eta)$. Thus by Lemma 8.2.1 (Nonswitching Extension) and Lemma 8.2.2 (Switching), we can find a switching extender ρ_0 of ρ_{-1} for $\pi_n^{k_n}$ if $n > 0$, and a nonswitching extender ρ_0 of ρ_{-1} for $\pi_n^{k_n} = \sigma^m$ if $n = 0$. By Lemma 9.2.1(i,ii) (PL Analysis), $\mathrm{PL}(\sigma^m, \xi, \rho_0) = \langle \pi_0^{k_0}, \dots, \pi_{n-1}^{k_{n-1}} \rangle$ and $\pi_{n-1}^{k_{n-1}}$ is $\lambda^{k_{n-1}}(\rho_0)$-free. Hence we can proceed as above by induction, replacing n with $n - j$ to obtain $\rho_n = \kappa$ satisfying the clauses of Definition 9.3.2. Note that the second clause of Definition 9.3.2(i) follows from Lemma 9.2.1(iii) (PL Analysis).

As each node of T^0 has only two outcomes, it follows from Lemma 8.2.1 (Nonswitching Extension) that switching immediate extensions of a node are unique. It is possible, however, that both extensions of a node of T^0 are nonswitching. If we always specify that we take the Σ outcome when we wish to take a nonswitching extension and both choices are acceptable, then the definition of the backtracking extender on T^0 is unique. ⊣

9.4. Backtracking and Links

We will prove two lemmas in this section. The first shows that the restoration of σ^m as a free node along the current path, requires switching all nodes in $\mathrm{PL}(\sigma^m, \xi, \eta)$. The second relates switching the outcome of the immediate antiderivative of a node in a PL set to switching the node in the PL set.

LEMMA 9.4.1 (Link Switching Lemma). *Fix $m \leq \ell$, $\sigma^m \in T^m$, $\sigma \in T^0$ and $\xi \subseteq \rho \in T^0$ such that σ^m is $\lambda^m(\xi)$-free and* $\mathrm{init}(\sigma^m, \xi) = \sigma$. *Then* $|\,\mathrm{PL}(\sigma^m, \xi, \rho)| = 1$ *iff both σ^m is $\lambda^m(\rho)$-free and* $\mathrm{init}(\sigma^m, \rho) = \sigma$.

PROOF. First suppose that $|\,\mathrm{PL}(\sigma^m, \xi, \rho)| = 1$. By Lemma 9.2.1(i,iii) (PL Analysis), σ^m is $\lambda^m(\rho)$-free and $\mathrm{init}(\sigma^m, \rho) = \sigma$.

Conversely, assume that σ^m is $\lambda^m(\rho)$-free and $\mathrm{init}(\sigma^m, \rho) = \sigma$. We assume that $|\,\mathrm{PL}(\sigma^m, \xi, \rho)| > 1$ and obtain a contradiction. Let $\mathrm{PL}(\sigma^m, \xi, \rho) = \langle \pi_i : i \leq n \rangle$ and for all $i \leq n$, fix k_i such that $\pi_i \in T^{k_i}$ and let μ_i be the initial derivative corresponding to π_i as defined in Definition 9.1.3. Let $j = \min\{k_i : 1 \leq i \leq n\}$, and fix the largest i such that $k_i = j$. By the choice of i and j, with notation as in Lemma 9.2.1 (PL Analysis), it follows from Lemma 9.2.1(v) (PL Analysis) that $\mu_i^j \subset \mu_{i-1}^j \subseteq \sigma^j$. As $[\mu_i^j, \pi_i^j]$ is a primary $\lambda^j(\rho)$-link, it follows from Lemma 8.3.3(iii) (Link Analysis) and the choice of i that $\sigma^{j+1} = \mathrm{up}^{j+1}(\sigma) \not\subseteq \lambda^{j+1}(\rho)$. But as σ^m is $\lambda^m(\rho)$-free and $\sigma = \mathrm{init}(\sigma^m, \rho)$, it follows from Lemmas 8.4.2 (Free Derivative) and 8.1.1(i) (Limit Path) that $\sigma^{j+1} \subseteq \lambda^{j+1}(\rho)$, yielding the desired contradiction. ⊣

The next lemma relates switches in an outcome of a node to switches of its principal derivative. It states that, in order to switch a node of T^k with Σ outcome when it has a fixed initial derivative, its principal derivative on T^{k-1} at that time must also be switched.

LEMMA 9.4.2 (Principal Switching Lemma). *Fix $k \leq \ell$, $\eta \subset \xi \in T^0$, $\pi^k \in T^k$ and $\pi^{k-1} \in T^{k-1}$ such that π^k is $\lambda^k(\eta)$-free and π^{k-1} is the principal derivative of π^k along $\lambda^{k-1}(\eta)$. Let $\widetilde{\pi} =$* $\mathrm{init}(\pi^k, \eta)$ *and $\widehat{\pi} =$* $\mathrm{init}(\pi^{k-1}, \eta)$. *Suppose that π^k is $\lambda^k(\xi)$-free and* $\mathrm{init}(\pi^k, \xi) = \widetilde{\pi}$, *and that no $\delta \in (\eta, \xi]$ switches π^k. Then π^{k-1} is the principal derivative of π^k along $\lambda^{k-1}(\xi)$, and $\widehat{\pi} =$* $\mathrm{init}(\pi^{k-1}, \xi)$. *Furthermore if $v^- = \xi$, v switches π^k and π^k has Σ outcome along $\lambda^k(\eta)$, then v switches π^{k-1}.*

PROOF. By Lemma 9.2.1(ii) (PL Analysis), π^k has the same outcome along $\lambda^k(\eta)$ and $\lambda^k(\xi)$, so π^{k-1} is the principal derivative of π^k along $\lambda^{k-1}(\xi)$ and has the same outcome along both $\lambda^{k-1}(\eta)$ and $\lambda^{k-1}(\xi)$. Thus $\widehat{\pi} = \mathrm{init}(\pi^{k-1}, \xi)$.

If π^k has Σ outcome along $\lambda^k(\eta)$, then π^k has the same outcome along $\lambda^k(\xi)$ and so π^{k-1} must have Π outcome along $\lambda^{k-1}(\xi)$. But then by (2.6.2), π^{k-1} is the only $\lambda^{k-1}(\xi)$-free derivative of π^k, and so if ν switches π^k then ν must also switch π^{k-1}. ⊣

9.5. Set Approximation

The progress of a construction through the trees of strategies will provide approximations to sets, based on a new weight function. These approximations will be related to the definition of the sets based on the true paths through the trees.

As we saw earlier, if $\eta^1 \in \{\lambda(\sigma) : \sigma \subset \Lambda^0\}$, then η^1 has a unique initial derivative $\subset \Lambda^0$; and if $\eta^2 \in \{\lambda^2(\sigma) : \sigma \subset \Lambda^0\}$, then η^2 has a unique initial derivative in $\{\lambda(\sigma) : \sigma \subset \Lambda^0\}$, and a unique initial derivative $\subset \Lambda^0$. This is no longer the case for $\eta^3 \in T^3$; η^3 may have many initial derivatives in $\{\lambda^2(\sigma) : \sigma \subset \Lambda^0\}$, and it is easiest to treat visits by the construction to η^3 differently when the visits are for different initial derivatives of η^3. Thus we define a new *weight function* as a function of two variables, a node on T^k for some k and an initial derivative of that node $\subset \Lambda^0$. We do not give a formal definition; rather, we leave it to the reader to show (the demonstration is routine) that we can define such a one-to-one function with domain $\mathbb{N} \times \mathbb{N}$ with the following property for $k \leq \ell$, $\sigma^k, \tau^k \in T^k$, and $\sigma, \tau \subset \Lambda^0$:

$$(\text{init}(\sigma^k, \sigma) = \sigma \text{ \& init}(\tau^k, \tau) = \tau \text{ \& } \sigma \subset \tau) \rightarrow \text{wt}(\sigma^k, \sigma) < \text{wt}(\tau^k, \tau). \tag{9.5.1}$$

For $k \leq \ell$ and $\eta \in T^0$, we also define

$$l_\eta^k = \max\{\text{wt}(\sigma^k, \sigma) + 1 : \sigma^k \in T^k \text{ \& } \sigma \subseteq \eta\}.$$

For convenience, we extend the domain of the function wt to $\{\langle \sigma^k, \sigma \rangle \in T^k \times T^0 : \sigma^k \subset \lambda^k(\eta)\}$ by setting $\text{wt}(\sigma^k, \eta) = \text{wt}(\sigma^k, \text{init}(\sigma^k, \eta))$.

We will be defining finite partial functions $\alpha_{a,\eta}$ with range $\{0, 1\}$ where a is a parameter coming from an outside set, and $\eta \subset \Lambda^0$. The domain of $\alpha_{a,\eta}$ will consist of those numbers x of the form $\text{wt}(\sigma^m, \sigma)$ for some $\sigma^m \in T^m$ encountered during the course of the construction, where $\sigma \subset \eta$ will be an initial derivative of σ^m. Each $\sigma^m \in T^m$ will have a *target set* $\text{TS}(\sigma^m)$ listing those parameters a for which $\alpha_{a,\eta}(\text{wt}(\sigma^m, \sigma))$ is eligible to take the value 1, and the map taking σ^m to $\text{TS}(\sigma^m)$ will be effective. With this background, the strings are now defined.

DEFINITION 9.5.1. Fix $m \leq \ell$, $\sigma^m \in T^m$ and $\sigma, \eta \in T^0$ such that $\sigma = \text{init}(\sigma^m, \sigma)$. Let a parameter a be given, and suppose the notion of *target set* with notation as above is defined. We define the string $\alpha_{a,\eta}$ as follows.

Case 1: $x = \mathrm{wt}(\sigma^m, \eta) < l^m_\eta$, $\alpha_{a,\eta^-}(x)\uparrow$ and $\sigma^m \subset \lambda^m(\eta)$. Define $\alpha_{a,\eta}(x) = 1$ if σ^m has Σ outcome along $\lambda^m(\eta)$ and $a \in \mathrm{TS}(\sigma^m)$, and define $\alpha_{a,\eta}(x) = 0$ otherwise.

Case 2: $x = \mathrm{wt}(\sigma^m, \eta)$, $\alpha_{a,\eta^-}(x)\downarrow$ and η switches σ^m. Define $\alpha_{a,\eta}(x) = 1 - \alpha_{a,\eta^-}(x)$ if $a \in \mathrm{TS}(\sigma^m)$, and define $\alpha_{a,\eta}(x) = 0$ otherwise.

Case 3: $x = \mathrm{wt}(\sigma^m, \sigma)$, $\sigma = \mathrm{init}(\sigma^m, \eta)$, $\alpha_{a,\eta^-}(x)\downarrow$ and η does not switch σ^m. Define $\alpha_{a,\eta}(x) = \alpha_{a,\eta^-}(x)$.

Case 4: Otherwise. Then $\alpha_{a,\eta}(x)\uparrow$.

For $i \leq \ell$, we also define $\alpha^{[i]}_{a,\eta}$ to be the restriction of $\alpha_{a,\eta}$ to those $x = \mathrm{wt}(\sigma^i, \sigma)$ such that $\sigma^i \in T^i$.

Before proceeding with an analysis of the effects of backtracking, we need to see when two strings on T^0 give rise to compatible sets.

DEFINITION 9.5.2. Fix $\xi \subset \eta \in T^0$ and a parameter a. We say that ξ and η are *a-compatible* if for all x, if $\alpha_{a,\xi}(x)\downarrow$ and $\alpha_{a,\eta}(x)\downarrow$, then $\alpha_{a,\xi}(x) = \alpha_{a,\eta}(x)$.

We now prove a lemma telling us that as we track a node σ^m at stages when it lies on the current path and has the same initial derivative σ, then looking at two consecutive such stages, if σ^m is not switched then it has the same type of outcome (Σ or Π) at both stages, and whenever σ^m is switched, the type of outcome is also switched.

LEMMA 9.5.3 (Faithfulness Lemma). *Fix a parameter* a, $m \leq \ell$, $\Lambda^0 \in T^0 \cup [T^0]$, $\sigma^m \in T^m$ *and* $\eta \subset \Lambda^0$ *such that* $\sigma^m \subset \lambda^m(\eta)$. *Let* $\sigma = \mathrm{init}(\sigma^m, \eta)$, *and fix the longest* $\nu \subseteq \eta$ *such that* $\lambda^m(\nu^-) \wedge \lambda^m(\nu) = \sigma^m$ *and* $\sigma = \mathrm{init}(\sigma^m, \nu)$. *Then*:

(i) σ^m *has* Σ *outcome along* $\lambda^m(\eta)$ *iff* σ^m *has* Σ *outcome along* $\lambda^m(\nu)$.
(ii) *If* $\sigma \subset \eta$, *then* $\alpha_{a,\eta}(\mathrm{wt}(\sigma^m, \sigma))\downarrow \in \{0, 1\}$. *Also*, $\alpha_{a,\eta}(\mathrm{wt}(\sigma^m, \sigma))\downarrow = 1$ *iff* σ^m *has* Σ *outcome along* $\lambda^m(\eta)$ *and* $a \in \mathrm{TS}(\sigma^m)$.

PROOF. We will be using results about the behavior of PL sequences relativized to $[0, m]$ in place of $[0, \ell]$. We proceed by induction on $|\eta|$. The lemma clearly holds for $\eta = \tau$, where $\tau^- = \sigma$. Note that if η switches σ^m, then by Lemma 8.5.1(iii) (Freeness Transmission), σ^m is both $\lambda^m(\eta^-)$-free and $\lambda^m(\eta)$-free, and σ^m has Σ outcome along $\lambda^m(\eta)$ iff σ^m has Π outcome along $\lambda^m(\eta^-)$. By Lemma 9.4.1 (Link Switching), if η switches σ^m and $\mathrm{init}(\sigma^m, \eta) = \sigma$, then $\mathrm{PL}(\sigma^m, \sigma, \eta) = \langle \sigma^m \rangle$. Furthermore, by Lemma 9.2.1(ii) (PL Analysis), if $\xi \subset \eta$, $\mathrm{PL}(\sigma^m, \sigma, \xi) = \mathrm{PL}(\sigma^m, \sigma, \eta) = \langle \sigma^m \rangle$, and $\mathrm{PL}(\sigma^m, \sigma, \rho) \neq \langle \sigma^m \rangle$ for all $\rho \in (\xi, \eta)$, then σ^m has the same outcome along $\lambda^m(\eta)$ as it had along $\lambda^m(\xi)$. (i) and (ii) now follow from Definition 9.5.1 and induction. ⊣

We now show that as long as ξ and η provide compatible path computations, the sets computed along those paths are compatible.

LEMMA 9.5.4 (Similarity Lemma). *Fix* $\eta_0 \subset \eta_1 \in T^0 \cup [T^0]$, *a parameter* a, *and* $m \leq \ell$ *such that* $\lambda^j(\eta_0) \subseteq \lambda^j(\eta_1)$ *for all* $j \leq m$. *Fix* $\sigma^m \in T^m$ *and* $\sigma \in T^0$

such that $\sigma = \text{init}(\sigma^m, \sigma)$, *let* $x = \text{wt}(\sigma^m, \sigma)$, *and assume that* $x < l^m_{\eta_0}$. *Then* $\alpha_{a,\eta_0}(x)\downarrow = \alpha_{a,\eta_1}(x)\downarrow$.

PROOF. We proceed by induction on $j \le m$, then by induction on $|\eta_1|$, and finally by induction on $|\eta_0|$. For $i \le 1$, fix the longest $\rho_i \subseteq \eta_i$ such that $\lambda^k(\rho_i^-) \wedge \lambda^k(\rho_i) = \sigma^m$ and $\text{init}(\sigma^m, \rho_i) = \sigma$. The lemma is clear for $m = 0$, as $\alpha^{[0]}_{a,\rho}$ is a finite initial segment of the characteristic function of $\emptyset$ of length l^0_ρ for all $\rho \in T^0$, and $l^0_{\eta_1} \ge l^0_{\eta_0}$.

Assume that $m > 0$. For all $i \le m$, let $\sigma^i = \text{up}^i(\sigma)$.

As $x < l^m_{\eta_0} < l^m_{\eta_1}$, $\alpha_{a,\eta_0}(x)\downarrow$ and $\alpha_{a,\eta_1}(x)\downarrow$. Let $\delta^m = \sigma^m \wedge \lambda^m(\eta_0)$. As $x < l^m_{\eta_0}$, we must have $\delta^m \subset \lambda^m(\eta_0)$, and so as $\lambda^m(\eta_0) \subseteq \lambda^m(\eta_1)$, $\delta^m = \sigma^m \wedge \lambda^m(\eta_1)$. If $\delta^m = \sigma^m$, then the lemma follows immediately from Lemma 9.5.3(ii) (Faithfulness); so assume that $\delta^m \subset \sigma^m$.

For $i \le 1$, fix the longest $\delta_i \subset \eta_i$ such that both $\text{up}^m(\delta_i^-) = \delta^m$ and $\lambda^m(\delta_i^-)|\lambda^m(\delta_i)$. As $\delta^m \subset \lambda^m(\eta_0) \subseteq \lambda^m(\eta_1)$, δ^m has the same outcome γ^m along both $\lambda^m(\eta_0)$ and $\lambda^m(\eta_1)$. Let $\gamma^j = \text{out}^j(\gamma^m)$ for all $j \le m$. By Remark 2.4.2(i) and Lemma 8.1.1(ii) (Limit Path), $\gamma^j = \lambda^j(\gamma^0) \subseteq \lambda^j(\eta_0), \lambda^j(\eta_1)$ for all $j \le m$, and by Lemma 8.5.3 (Crossover), if j is the largest number such that $\sigma^i \subset \gamma^i$ for all $i \le j$, then σ^j is properly restrained by a primary γ^j-link whose last element is $(\gamma^j)^-$. Thus in particular, $x < l^m_{\gamma^0}$. If $\gamma^0 \subset \eta_0$, then by induction, $\alpha_{a,\eta_0}(x) = \alpha_{a,\gamma^0}(x) = \alpha_{a,\eta_1}(x)$, so the lemma follows.

It remains to consider the case wherein $\eta_0 = \gamma^0$. If $j > 0$, then by relativizing the proof to start on T^1 rather than on T^0, we see that the hypotheses are similar but the induction level is smaller by 1, so the lemma follows by induction in this case. But if $j = 0$, then it follows from Remark 2.4.2(ii) that $\delta_0 = \delta_1$ and hence $\rho_0 = \rho_1$. Thus by Lemma 9.5.3(i,ii) (Faithfulness), $\alpha_{a,\eta_0}(x) = \alpha_{a,\gamma^0}(x) = \alpha_{a,\rho_0}(x) = \alpha_{a,\eta_1}(x)$, so the lemma follows in this case as well. $\dashv$

The next lemma will be used to that the backtracking procedure enables us to correct axioms.

LEMMA 9.5.5 (Weight Comparison Lemma). *Let* $m \le \ell$, $\sigma^m \in T^m$ *and* $\sigma \subseteq \xi \subseteq \eta \in T^0$ *be given such that* $\text{init}(\sigma^m, \xi) = \sigma$ *and* $\text{lev}(\sigma^m) = m$. *Let* $\text{PL}(\sigma^m, \xi, \eta) = \langle \pi_i^{k_i} : i \le n\rangle$ *where* $\pi_i^{k_i} \in T^{k_i}$. *For each* $i \in [1, n]$, *fix the shortest* $\rho_i \subseteq \eta$ *such that* $|\text{PL}(\sigma^m, \xi, \delta)| \ge i + 1$ *for all* $\delta \in [\rho_i, \eta]$, *and let* $\pi_i = \text{init}(\text{up}(\pi_i^{k_i}), \rho_i)$. *Then for all* $i \in [1, n]$, $\text{wt}(\text{up}(\pi_i^{k_i}), \pi_i) \le l^{k_i+1}_\sigma$.

PROOF. Fix $i \in [1, n]$, and let $\mu_i = \text{init}(\pi_i^{k_i}, \rho_i)$. It is easy to see that $\pi_i \subseteq \mu_i$. By Lemma 9.2.1(v) (PL Analysis), $\mu_i \subseteq \sigma$. As $\text{lev}(\sigma^m) = m$, Definition 9.1.3 rules out the possibility that $\mu_i = \sigma$. Thus $\pi_i \subseteq \mu_i \subset \sigma$, so by (9.5.1), $\text{wt}(\text{up}(\pi_i^{k_i}), \pi_i) \le l^{k_i+1}_\sigma$. $\dashv$

9.6. Constraint

Some constructions, upon reaching certain nodes $\eta \in T^0$, will require that we pass from η to the backtracking extender κ of η for some pair $\langle \sigma^m, \sigma \rangle$. When this occurs, we refer to the passage from η to κ as the *backtracking process*. During this process certain nodes will become *constrained*. It is the purpose of this section to define the constrained nodes, and prove a lemma about the nature of constraint within a construction.

Every constrained node must have the following property.

DEFINITION 9.6.1. Fix $\delta \in T^0$ and $\delta^k \in T^k$. We say that δ^k is δ-*expansionary* if $\text{up}^k((\text{out}^j(\delta^k))^-) = (\delta^k)^-$ for all $j \leq k$. (In other words, for all $j \leq k$, if $\delta^j = \text{out}^j(\delta^k)$ and $\pi^j = (\delta^j)^-$, then π^j is the principal derivative of π^k along $\lambda^j(\delta)$.)

When backtracking is implemented to pass from η to κ, outcomes of the nodes encountered along the way are forced independently of the truth value of the directing sentences of those nodes. The expansionary nodes encountered become constrained, as per our next definition.

DEFINITION 9.6.2. Suppose that a construction implements backtracking at η from ξ for $\langle \sigma^m, \sigma \rangle$. Let κ be the backtracking extender of η from ξ for $\langle \sigma^m, \sigma \rangle$, and fix m such that $\sigma^m \in T^m$. We say that δ^k is *constrained* (*for* $\langle \sigma^m, \sigma \rangle$) if $\text{out}^0(\delta^k) \in (\eta, \kappa]$. (Thus the nodes constrained for $\langle \sigma^m, \sigma \rangle$ are those that are δ-expansionary for some $\delta \in (\eta, \kappa]$.)

Note that, with notation as in the preceding definition, all nodes in $(\eta, \kappa]$ are constrained.

In order for the backtracking process to be well-behaved, we need to impose restrictions on the situations in which it is allowed to be implemented. When these restrictions are obeyed, we call the backtracking *normal*.

DEFINITION 9.6.3. We say that a construction with true path Λ^0 implements *normal backtracking* if all instances of backtracking within the construction satisfy the following conditions. Suppose that we implement backtracking starting at η for $\langle \sigma^m, \sigma \rangle$ from ξ, that κ is the corresponding backtracking extender for η, and that ρ is the immediate successor of κ along Λ^0. Then:

(i) ρ switches σ^m and $\lambda^m(\rho)$ is ρ-expansionary.
(ii) σ^m has Σ outcome along $\lambda^m(\rho)$.
(iii) Neither the immediate successor of σ along Λ^0 nor ρ is constrained.
(iv) If $\delta \subset \zeta$, either $\lambda^k(\delta)$ is constrained or $\text{out}^0(\lambda^k(\delta))^-$ is a backtracking extender, and $(\lambda^k(\delta))^-$ is switched by ζ, then ζ is constrained.

The next lemma allows us to show that the construction is faithfully executed even though action at constrained nodes is restricted.

LEMMA 9.6.4 (Constraint Lemma). *Suppose that a given construction implements normal backtracking. Then if* ζ^k *is constrained,* $\mathrm{out}^0(\zeta^k) \subseteq \xi$ *and* ξ *is not constrained, then* $(\zeta^k)^-$ *is not* $\lambda^k(\xi)$*-free.*

PROOF. We begin by describing the configuration of constrained nodes on T^0. Suppose that the construction passes from η_0 to its backtracking extender κ_0 for $\langle \sigma_0^{m_0}, \sigma_0 \rangle$, and that no antiderivatives of constrained nodes or backtracking extenders are switched by nodes in $(\eta_0, \kappa_0]$. Let ρ_0 be the immediate successor of κ_0 along Λ^0. Then all nodes in $(\eta_0, \kappa_0]$ are constrained and by clause (iii) of Definition 9.6.3, ρ_0 is not constrained. Furthermore, by clauses (i)–(iii) of Definition 9.6.3, $[\sigma_0, \kappa_0]$ is a ρ_0-link.

By clause (iv) of Definition 9.6.3, σ_0^m can only be switched by a node extending ρ_0 when another backtracking process is implemented. As, by (2.6.4), only free nodes can be switched, it follows from Lemma 8.3.3(i) (Link Analysis) that no node in $[\eta_0, \kappa_0)$ can become free at a node extending ρ_0 unless the construction passes from some $\eta_1 \supseteq \rho_1$ to its backtracking extender κ_1 for some $\langle \sigma_1^{m_1}, \sigma_1 \rangle$, and in the course of the backtracking, some $\delta \in (\eta_1, \kappa_1]$ switches $\mathrm{up}^{m_0-1}(\kappa_0)$. In this case, if ρ_1 is the immediate successor of κ_1 along Λ^0 and $\widetilde{\kappa}_0 = \delta^-$, then we will have a ρ_1-link from κ_0 to $\widetilde{\kappa}_0$, and a ρ_1-link from σ_1 to κ_1, and by clause (iii) of Definition 9.6.3, $\sigma_0 \subset \sigma_1 \subset \eta_0$.

Continuing in this way, we can assume inductively that we have a configuration of an enclosed linked chain, i.e., nodes

$$\sigma_n \subset \eta_0 \subset \kappa_0 \subset \eta_1 \subseteq \widetilde{\kappa}_0 \subset \kappa_1 \subseteq \cdots \subseteq \eta_n \subseteq \widetilde{\kappa}_{n-1} \subset \kappa_n \subset \rho_n$$

such that each interval $(\eta_j, \kappa_j]$ consists of constrained nodes, and $[\kappa_j, \widetilde{\kappa}_j]$ for $j < n$ and $[\sigma_n, \kappa_n]$ are ρ_n-links. If another implementation of backtracking, say from η_{n+1} to κ_{n+1} for $\langle \sigma_{n+1}^{m_{n+1}}, \sigma_{n+1} \rangle$ revises this configuration, then by clause (iii) of Definition 9.6.3, either $\sigma_n \subset \sigma_{n+1} \subset \eta_0$ in which case one more link is added to the configuration, or $\sigma_{n+1} \in [\rho_j, \eta_{j+1})$ for some j. In the latter case, we will have a configuration as described above starting at η_{j+1}. Furthermore, in order to make $\sigma_{n+1}^{m_{n+1}}$ free, we will have to switch $\mathrm{up}^{m_j-2}(\kappa_j)$, thereby reestablishing a shorter such configuration from η_0 to ρ_j.

The lemma now follows from this description, as no constrained node can be free at a non-constrained node. $\dashv$

9.7. REA Sets

We will show, in this section, that the sets constructed in Chapter 10 lie in the REA hierarchy. These are the sets for which the partial functions $\alpha_{a,\eta}$ provide an approximation, as we now describe.

DEFINITION 9.7.1. We define $A_a(x)$ to be $\alpha_{a,\eta}(x)$ for the shortest η such that $\lambda^j(\eta) \subset \Lambda^j$ for all $j \leq \ell$ and $x \in \mathrm{dom}(\alpha_{a,\eta})$. If $i \leq \ell$, then $A_a^{[i]}(x) \downarrow$ if $x \in \mathrm{dom}(\alpha_{a,\eta}^{[i]})$, and in this case, $A_a^{[i]}(x) \downarrow = A_a(x)$.

Note that by Lemma 9.5.4 (Similarity), we can choose any $\xi \supset \eta$ such that $\lambda^j(\xi) \subset \Lambda^j$ for all $j \leq \ell$ in the preceding definition without changing the value of $A_a(x)$.

Our next lemma allows us to show that the sets just defined lie in the REA hierarchy.

DEFINITION 9.7.2. A set A is said to be 0-*REA* if it is computable. A is $m+1$-*REA* if there is an m-REA set $B \leq_T A$ such that A is computably enumerable in B.

LEMMA 9.7.3 (REA-ness Lemma). *Let a parameter a and $m \leq \ell$ be given such that* $\mathrm{lev}(a) = m$. *Then A_a is m-REA.*

PROOF. We proceed by induction on m. Clearly, $A_a^{[0]}$ is 0-REA for all $a \in J$. Suppose that the lemma holds for $m-1$. Then $A_a^{[j]}$ is j-REA for all $j \leq m-1$, and $\bigcup\{A_a^{[j]} : j \leq m-1\}$ is $(m-1)$-REA.

Claim: $A_a^{[m]}(x) = 1$ iff there are $\sigma^m \in T^m$, $\sigma, \widetilde{\sigma}, \eta \subset \Lambda^0$, and $\sigma^{m-1} \subset \lambda^{m-1}(\eta)$ such that the following conditions hold:

(i) $\sigma = \mathrm{init}(\sigma^m, \eta)$ and $\widetilde{\sigma} = \mathrm{init}(\sigma^{m-1}, \eta)$.
(ii) $x = \mathrm{wt}(\sigma^m, \sigma)$.
(iii) $\mathrm{up}(\sigma^{m-1}) = \sigma^m$.
(iv) σ^{m-1} has Π outcome along $\lambda^{m-1}(\eta)$.
(v) If $y = \mathrm{wt}(\sigma^{m-1}, \widetilde{\sigma})$, then $\alpha_{a,\eta}^{[m-1]}(y) = A_a^{[m-1]}(y)$.
(vi) $a \in \mathrm{TS}(\sigma^m)$.

It follows from the claim that $A_a^{[m]}$ is computably enumerable in $A_a^{[m-1]}$, so lies in the REA hierarchy. Thus it suffices to prove the claim.

Proof of Claim. First suppose that (i)–(vi) hold. If $\tau^\ell \subset \Lambda^\ell$, $|\tau| > 0$ and $\tau = \mathrm{out}^0(\tau^\ell)$ then by Remark 2.4.2(i) τ is Λ^0-true. Hence by Lemma 8.1.1(iv) (Limit Path), we may pass to the shortest ρ such that $\eta \subseteq \rho \subset \Lambda^0$ and ρ is Λ^0-true. By Lemma 9.5.4 (Similarity), $\alpha_{a,\rho}^{[i]} = A_a^{[i]} \upharpoonright l_\rho^i$ for all $i \leq \ell$.

Fix the longest $\nu \subseteq \rho$ such that σ^{m-1} is $\lambda^{m-1}(\nu)$-free and $\mathrm{init}(\sigma^m, \nu) = \mathrm{init}(\sigma^m, \eta)$; the existence of ν follows from (i), Lemma 8.1.1(i) (Limit Path), and the definition of links. By (2.6.4) and Lemma 8.4.2(i) (Free Derivative), σ^{m-1} is not switched at any $\beta \in (\nu, \rho]$ such that $\mathrm{init}(\sigma^m, \beta) = \sigma$, so by Definition 9.5.2, $\alpha_{a,\nu}^{[m-1]}(y) = \alpha_{a,\rho}^{[m-1]}(y)$. Furthermore, by (iv), σ^{m-1} has Π outcome along $\lambda^{m-1}(\nu)$. Hence by (v) and Definition 9.1.3,

$$\alpha_{a,\nu}^{[m-1]}(y) = \alpha_{a,\rho}^{[m-1]}(y) = A_a^{[m-1]}(y) = \alpha_{a,\eta}^{[m-1]}(y) = 0.$$

Now as σ^{m-1} has Π outcome along $\lambda^{m-1}(\nu)$, σ^{m-1} is the principal derivative of σ^m along $\lambda^{m-1}(\nu)$, and by Remark 2.4.2(i), σ^m has Σ outcome along $\lambda^m(\nu)$. By

Lemma 9.4.2 (Principal Switching) and Definition 9.1.3, $\alpha_{a,v}^{[m]}(x) = \alpha_{a,\rho}^{[m]}(x) = 1$. As $\alpha_{a,\rho}^{[m]} \subset A_a^{[m]}$, $A_a^{[m]}(x) = 1$.

Conversely, suppose that $A_a^{[m]}(x) = 1$. Then there are $\sigma^m \in T^m$ and $\sigma \subset \eta \subset \Lambda^0$ such that $\sigma = \text{init}(\sigma^m, \eta)$, $x = \text{wt}(\sigma^m, \sigma)$, $\text{up}^m(\eta^-) = \sigma^m$, $a \in \text{TS}(\sigma^m)$, and σ^m has Σ outcome along $\lambda^m(\eta)$. Let σ^{m-1} be the principal derivative of σ^m along $\lambda^{m-1}(\eta)$, and let $\widetilde{\sigma} = \text{init}(\sigma^{m-1}, \eta)$. By Remark 2.4.2(i), σ^{m-1} has Π outcome along $\lambda^{m-1}(\eta)$. Thus if $y = \text{wt}(\sigma^{m-1}, \widetilde{\sigma})$, then by Lemma 9.5.3(ii) (Faithfulness), $\alpha_{a,\eta}^{[m-1]}(y) = A_a^{[m-1]}(y) = 0$. We conclude that (i)–(vi) hold. ⊣

9.8. Level-Specific Backtracking

We now prove several lemmas that were used in backtracking constructions at levels ≤ 3.

LEMMA 9.8.1. *Fix $\eta^k \subset \Lambda^k \in [T^k]$ such that η^k is Λ^k-free, and suppose that a construction implements only normal backtracking. Then $|\{\rho^k \subset \Lambda^k : \eta^k$ is ρ^k-free & ρ^k is not constrained$\}| = \infty$.*

PROOF. Immediate from Lemmas 8.7.2(iv) (Σ_1-Analysis) and 9.6.4 (Constraint). ⊣

LEMMA 9.8.2. *Suppose that a construction takes place only on trees T^i for $i \leq 3$. and implements only normal backtracking. Then for any $\eta \subset \Lambda^0$, there is a $\rho \in [\eta, \Lambda^0)$ such that ρ^- is not constrained.*

PROOF. Immediate from Lemmas 8.7.2(iv) (Σ_1-Analysis) and 9.6.4 (Constraint). ⊣

Chapter 10

HIGHER-LEVEL CONSTRUCTIONS

The framework we are using was originally developed by Lempp and Lerman [13] to handle requirements that must be placed on high level trees. We will use it in this chapter to prove Lerman's [16] result that the existential theory of the jump poset of degrees with least element is decidable.

10.1. Jump Posets with Least Element

One of the central themes of degree theory is to try to obtain decidability and undecidability information about elementary theories of structures naturally induced from the relation *is computable from*, and of natural fragments of those elementary theories. We present a proof of such a result here, showing that the existential theory of the jump poset $\langle \boldsymbol{D}, \leq, \boldsymbol{0}, ' \rangle$ of the Turing degrees with least element is decidable. The decision procedure centers on a solution to a related problem of an order-theoretic nature; obtaining a characterization of the jump posets with least element that can be embedded into the jump poset of Turing degrees with least element $\boldsymbol{0}$. We show that all jump posets with least element can be so embedded.

DEFINITION 10.1.1. A *jump poset with least element* is a structure $\mathcal{P} = \langle P, \leq, 0, ' \rangle$ such that $\langle P, \leq \rangle$ is a poset with least element 0, and $'$ is a *strictly monotone* and *increasing* function, i.e., for all $a, b \in P$

(i) $a < a'$.
(ii) $a \leq b \rightarrow a' \leq b'$.

$\mathcal{P}$ is said to have *finite support* if there is a natural number ℓ such that $S = \{a \in P : a \leq 0^{(\ell)}\}$ is finite and for all $a \in P - S$, $a = 0^{(m)}$ for some natural number $m > \ell$; the smallest such ℓ is called the *support level* of $\mathcal{P}$. A *finite partial structure* $\mathcal{Q}$ in this language consists of:

(iii) A finite set P.
(iv) Finitely many equalities of the form $t(a_1, \ldots, a_n) = c$ where $t(a_1, \ldots, a_n)$ is a term in the language of jump posets with least element having no free variables and $a_1, \ldots, a_n, c \in P$.

(v) Finitely many inequalities of the form $c \not\leq d$ where $c, d \in P$.

A finite poset $\mathcal{Q}$ is *good* if its diagram is consistent with the axioms for jump posets with least element.

The order-theoretic theorem is:

THEOREM 10.1.2. *If $\mathcal{P}$ is a good finite jump poset with least element and finite support, then $\mathcal{P}$ can be embedded into $\langle \boldsymbol{D}, \leq, \boldsymbol{0}, ' \rangle$ (preserving least element). Moreover, there is such an embedding into the REA degrees $\mathbf{D_{REA}}$ and hence into arithmetical degrees $\mathbf{D_{arith}}$.*

An existential sentence in the language of jump posets with least element can easily be seen to be equivalent to a statement asserting that one of finitely many given finite partial structures in the language of jump posets with least element can be embedded into $\langle \boldsymbol{D}, \leq, \boldsymbol{0}, ' \rangle$. It is easy to see that any good finite partial jump poset with least element can be embedded into a jump poset with least element and finite support. Hence the following corollary is immediate.

COROLLARY 10.1.3. *(i) The existential theory of $\langle \boldsymbol{D}, \leq, \boldsymbol{0}, ' \rangle$ is decidable.*
(ii) The existential theory of $\langle \mathbf{D_{arith}}, \leq, \boldsymbol{0}, ' \rangle$ is decidable.
(iii) The existential theory of $\langle \mathbf{D_{REA}}, \leq, \boldsymbol{0}, ' \rangle$ is decidable.

10.2. The Requirements

For the remainder of this chapter, fix a jump poset $\mathcal{P} = \langle P, \leq, 0, ' \rangle$ with finite support having level ℓ. The embedding will map $a \in P$ to A_a.

DEFINITION 10.2.1. Each element $a \in P$ will have a *level*, denoted as $\text{lev}(a)$, and defined by $\text{lev}(a) = n$ if n is the smallest natural number such that $a \leq 0^{(n)}$. In addition, each element $a \in P$ will have a *height*, denoted as $\text{ht}(a)$, and defined by $\text{ht}(a) = n$ if n is the largest natural number such that $0^{(n)} \leq a$. For $i \leq \ell$, let $P^{[i]} = \{a \in P : \text{lev}(a) = i\}$, $P^{[\geq i]} = \cup\{P^{[k]} : k \in [i, \ell]\}$; $P^{[>i]}$, $P^{[<i]}$ and $P^{[\leq i]}$ are defined analogously.

Nodes of T^k, and requirements assigned to these nodes, will be of one of two types. Requirements of type 0 ensure the preservation of inequality of elements of P by the embedding into the degrees, and those of type 1 ensure that relations $a' \leq b$ are preserved by the embedding for $a, b \in P$. The preservation of $b \leq a'$ and $\leq$ by the embedding will be a direct result of the construction, but we will establish requirements of types 2 and 3 to delineate what needs to be verified.

We will take action to ensure the satisfaction of requirements. Thus we will associate a subset $\text{TS}(R, i)$ of $P^{[\geq i]}$ with a requirement R for each $i \leq \ell$, called the *target set* for R of level i. This set will consist of those $a \in P^{[\geq i]}$ for which weights of elements of T^i to which R is assigned are to enter $A_a^{[i]}$. For each requirement R, we will choose a $c \in P$ such that $\text{TS}(R, i) = \text{TS}(c, i)$ (defined below) for all $i \leq \ell$. If a requirement R is assigned to a node $\eta^k \in T^k$, then we define $\mathrm{TS}(\eta^k) = \mathrm{TS}(R, k)$.

Target sets for requirements and nodes of trees will be generated by elements of P as follows.

DEFINITION 10.2.2. Fix $a \in P$ and $i \leq \ell$. We define $\mathrm{TS}(a, i) = \{b \in P^{[\geq i]} : b^{\mathrm{ht}(a)-i+1} \not\leq a\}$ (if $i > \mathrm{ht}(a)+1$, then we specify that b satisfies the condition). Note that $\mathrm{TS}(a, i)$ is upwards closed in $P^{[\geq i]}$.

The following lemma delineates properties of target sets that will be used later.

LEMMA 10.2.3 (Target Set Lemma). *Fix* $a, b, c \in P$ *and* $i \leq \ell$. *Then*:

(i) $(a \leq b \;\&\; a \in \mathrm{TS}(c, i)) \rightarrow b \in \mathrm{TS}(c, i)$.

(ii) $(i > 0 \;\&\; b \leq a' \;\&\; b \in \mathrm{TS}(c, i)) \rightarrow a \in \mathrm{TS}(c, i-1)$.

(iii) $(0 < i \leq \mathrm{lev}(b) \;\&\; a' \leq b \;\&\; a \in \mathrm{TS}(c, i-1)) \rightarrow b \in \mathrm{TS}(c, i)$.

PROOF. Let $h = \mathrm{ht}(c)$.

(i): As $a < b$, $a^{(h-i+1)} \not\leq c$ implies, by Definition 10.1.1(ii), that $b^{h-i+1} \not\leq c$.

(ii): Suppose that $b \leq a'$. By hypothesis, $b^{(h-i+1)} \not\leq c$. Now by Definition 10.1.1(ii),

$$b^{(h-i+1)} \leq (a')^{(h-i+1)} = a^{(h-i+2)} = a^{(h-(i-1)+1)},$$

so $a \in \mathrm{TS}(c, i-1)$.

(iii): Suppose that $a' \leq b$. By hypothesis, $a^{(h-(i-1)+1)} \not\leq c$. But by Definition 10.1.1(ii),

$$a^{(h-(i-1)+1)} = (a')^{(h-i+1)} \leq b^{(h-i+1)},$$

so $b^{(h-i+1)} \not\leq c$. ⊣

We now introduce the requirements to be satisfied by the construction. There will be four types of requirements, and we define $\mathrm{tp}(R)$ to be the type of the requirement R.

DEFINITION 10.2.4. *Incomparability requirements* will ensure the satisfaction of the relation $A_b \not\leq_T A_a$, where $a, b \in P$ and $b \not\leq a$. They are *type* 0, *level* $k = \mathrm{ht}(a) + 1$ requirements; such a requirement $R = R^{0,k}_{e,b,a}$ is introduced for each $e \in \mathbb{N}$ stating that $\exists x(A_b(x) \neq \Phi_e(A_a; x))$. For $i \leq \ell$, we set $\mathrm{TS}(R, i) = \mathrm{TS}(a, i)$.

There are two different types of requirements dealing with jump comparability relations.

DEFINITION 10.2.5. *Jump comparability requirements of type* 1 correspond to relations $a' \leq b$, where $a, b \in P$ and $a' = b$. They are *level* $k = \mathrm{ht}(a) + 1$ requirements. Such a requirement $R = R^{1,k}_{e,b,a}$ is introduced for each $e \in \mathbb{N}$ stating that $\Delta^{1,k}_{b,a}(A_b, e) = 1$ if $\Phi_e(A_a; e)\downarrow$, and $\Delta^{1,k}_{b,a}(A_b, e) = 0$ if $\Phi_e(A_a; e)\uparrow$, where $\Delta^{1,k}_{b,a}$ is a computable partial functional that will be defined during the construction for the purpose of satisfying all such requirements for fixed a and b. For $i \leq \ell$, we $\mathrm{TS}(R, i) = \mathrm{TS}(a, i)$.

DEFINITION 10.2.6. *Jump comparability requirements of type* 2 correspond to relations $b \leq a'$, where $a, b \in P$ and $b = a'$. Such a requirement $R = R^{2,k}_{e,b,a}$ is introduced for each $e \in \mathbb{N}$ stating that $\Delta^{2,k}_{b,a}(A_a; e, s) = 1$ for cofinitely many s if $e \in A_b$, and $\Delta^{2,k}_{b,a}(A_a; e, s) = 0$ for cofinitely many s if $e \notin A_b$, where $\Delta^{2,k}_{b,a}$ is a computable partial functional that will be defined for the purpose of satisfying all such requirements for fixed a and b.

DEFINITION 10.2.7. *Comparability requirements of type* 3 assert that $A_a \leq_T A_b$ whenever $a, b, \in P$ and $a \leq b$.

LEMMA 10.2.8 (Requirement Lemma). *Assume that we have a collection of sets $\{A_a : a \in P\}$ for which the above requirements are satisfied. Then the map from $\mathcal{P}$ to $\mathcal{D}$ sending $a \in P$ to the degree of A_a is a jump poset embedding preserving least element.*

PROOF. Requirements of types 0 and 3 combine to show that the map is an order-preserving poset embedding. We note that $A_0 = \emptyset$, so the map also preserves least element. Requirements of types 1 and 2 combine to show that the map preserves the jump operation. ⊣

10.2.1. T^ℓ-Analysis. The standard initial assignment process is satisfied. Definition 2.7.2(i) Follows easily from Lemma 2.7.4 (Standard Initial Assignment). Hence it suffices to show that all requirements are satisfied.

The standard initial sentence specification for $\eta^\ell \in T^\ell$ is followed, with action for $\eta^\ell \in T^\ell$ supported at $\{\rho^\ell \subset \eta^\ell : \eta^\ell \text{ is } \rho^\ell\text{-free}\}$. Clause (i) of Definition 2.10.1 follows from the analysis for basic modules.

10.2.2. T^i-Analysis, $i < \ell$. The standard derived assignment is followed. Clauses (i)–(iii) of Definition 2.7.7 follow from Lemma 2.12.2 (Standard Derived Assignment). Sentence decomposition will be defined later, and the conditions relating to sentence decomposition will be checked at that time. Action for $\eta^i \in T^i$ will be supported at $\{\rho^i \supset \eta^i : \eta^i \text{ is } \rho^i\text{-free and } \rho^i \text{ is not constrained}\}$. By Lemma 9.8.1, the construction has infinite support on T^i.

10.3. The Construction

In this section, we will present the construction and prove some technical lemmas. The proof of the satisfaction of requirements is deferred to the next section.

We now indicate how to specify the sentence that generates the action for a requirement assigned to a given node. We will show that for requirements assigned to nodes with the property that all their antiderivatives lie on the true paths (through the appropriate trees) determined by the construction, the truth of the sentence locally for an oracle $\alpha_{a,\eta}$ ensures the truth of the sentence for the oracle A_a.

DEFINITION 10.3.1 (Directing Sentences). Fix $\xi \in T^0$. Let $R_\xi = R^{j,k}_{e,b,a}$, and note that $k = \mathrm{lev}(R)$. Let $\xi^k = \mathrm{up}^k(\xi)$, and let σ be the initial derivative of ξ^k along ξ. We will assign a *directing sentence* M_ξ to ξ as follows. If $j = 0$, let M_ξ be $\Phi_e(\alpha_{a,\xi}; \mathrm{wt}(\xi^k, \sigma))[\mathrm{wt}(\xi)] \downarrow = 0$; and if $j = 1$, let M_ξ be the sentence $\Phi_e(\alpha_{a,\xi}; e)[\mathrm{wt}(\xi)] \downarrow$. We also define $M_{\xi/\eta}$ to be $\Phi_e(\alpha_{a,\eta}; \mathrm{wt}(\xi^k, \sigma))[\mathrm{wt}(\eta)] \downarrow = 0$ if $j = 0$, and $\Phi_e(\alpha_{a,\eta}; e)[\mathrm{wt}(\eta)] \downarrow$. (Thus $M_{\xi/\eta}$ evaluates the truth of the directing sentence for ξ at stage η.)

The action taken for a requirement associated with $\eta \in T^0$ that is not constrained depends on the truth of the directing sentences $M_{\xi/\eta}$ for $\xi \subseteq \eta$. If a node η is associated with the type 1 requirement $R_\eta = R^{1,k}_{e,b,a}$, and action specified for the requirement involves declaring axioms for the functional $\Delta^{1,k}_{b,a}(A_b)$, then we use the notation Δ_η to denote $\Delta^{1,k}_{b,a}$. We now describe the situation in which the construction must act non-trivially. The conditions enumerated below describe the situation in which we are forced to take action in order to satisfy a requirement assigned to a specified node.

DEFINITION 10.3.2. Fix $\xi \subseteq \eta \in T^0$, and let $R_\eta = R^{j,k}_{e,b,a}$. Let $\xi^k = \mathrm{up}^k(\xi)$ and $\eta^k = \lambda^k(\eta)$, and let $x = \mathrm{wt}(\xi^k, \mathrm{init}(\xi^k, \xi))$ if $j = 0$ and set $x = e$ if $j = 1$. We say that η *requires attention for* ξ if $\mathrm{lev}(\xi) = k$, neither ξ nor η^- is constrained, $M_{\xi/\eta}$ is true, either $\alpha_{b,\eta}(x) \uparrow$ or $\alpha_{b,\eta}(x) = 0$, if $j = 0$ then $\lambda^j(\xi)$ is $\lambda^j(\eta)$-true for all $j \leq \ell$, and if $j = 1$ then $\alpha_{b,\xi}$ and $\alpha_{b,\eta}$ are compatible.

10.3.1. The Construction. We define a path $\Lambda^0 \in [T^0]$ by induction on $|\eta|$ for $\eta \subset \Lambda^0$. We begin by specifying that $\langle\,\rangle \subset \Lambda^0$. We determine an extension τ of η, and will declare axioms for requirements of type 1 in Step 2. This part of the construction is called *stage* η.

Step 1: There are two cases.

Case 1: η requires attention for some ξ. Fix the shortest such ξ, and let $R_\xi = R^{j,k}_{e,b,a}$, $k = \mathrm{lev}(\xi)$, $\xi^k = \mathrm{up}^k(\xi)$, and $\widetilde{\xi} = \mathrm{init}(\xi^k, \xi)$. Let κ be the backtracking extender of η for $\langle \xi^k, \widetilde{\xi}\rangle$ from ξ; if no such κ exists, then the construction is terminated. Now fix τ such that $\tau^- = \kappa$ and ξ^k has Σ outcome along $\lambda^k(\tau)$, and specify that $\tau \subset \Lambda^0$.

Case 2: Otherwise. Let $R_\eta = R^{j,k}_{e,b,a}$. We define an extension $\tau \subset \Lambda^0$ of η such that $\tau^- = \eta$ as follows. $\tau = \eta^\frown\langle\infty\rangle$ if either $\mathrm{ht}(\eta)$ is odd and M_η is true or if $\mathrm{ht}(\eta)$ is even and M_η is false, and $\tau = \eta^\frown\langle 0\rangle$ otherwise. Specify that $\tau \subset \Lambda^0$.

Step 2: We proceed with this step only if either Case 1 is followed and $\text{tp}(\xi) = 1$, or if Case 2 is followed and $\text{tp}(\eta) = 1$. We *declare an axiom* $\Delta_\tau(\alpha_{b,\tau}; e) = 1$ if M_{τ^-} is true, and set $\Delta_\tau(\alpha_{b,\tau}; e) = 0$ otherwise.

The construction is now complete. For all $i \leq \ell$, let $\Lambda^i = \lambda^i(\Lambda^0)$.

10.4. Proof

The following lemma is used to show that the implementations of backtracking in the construction succeed.

LEMMA 10.4.1 (Backtracking Implementation Lemma). *Suppose that the construction specifies that we pass to the backtracking extender of* v *for* $\langle \sigma^m, \sigma \rangle$ *from* ξ. *Then such a backtracking extender* κ *exists. Furthermore, if* R_{σ^m} *is* $R^{j,k}_{e,a,b}$, *then* $\alpha_{a,v} \subseteq \alpha_{a,\kappa}$.

PROOF. First suppose that $j = 0$. Then $\text{up}^k(\sigma)$ is $\lambda^k(v)$-true for all $k \leq \ell$, so by Lemma 9.4.1 (Link Switching), $|\text{PL}(\sigma^m, \xi, v)| = 1$. Hence by Lemma 9.3.3 (Backtracking), a unique backtracking extender κ exists. Furthermore, no links are switched in the passage from v to κ, so $\alpha_{a,v} \subseteq \alpha_{a,\kappa}$.

Now suppose that $j = 1$. By Lemma 9.3.3 (Backtracking), κ will exist unless some node $\pi \in \text{PL}(\sigma^m, \xi, v)$ lies in T^0. Now as backtracking is specified, we must have $\alpha_{b,\xi} \subset \alpha_{b,v}$; but as $\pi \in \text{PL}(\sigma^m, \xi, v)$, $\pi^1 = \text{up}(\pi) \subset \lambda(\xi) \wedge \lambda(v)$. Assume that this is the case in order to obtain a contradiction. Thus by Lemma 9.5.5 (Weight Comparison), $\text{wt}(\pi^1, \text{init}(\pi^1, \pi)) \in \text{dom}(\alpha_{b,\xi})$. Let $\text{TS}(\pi^1) = \text{TS}(c, 1)$. Now $b \in \text{TS}(c, 1)$ iff $b^{(\text{ht}(c))} \not\leq c$; but as $b \geq a'$, $b \geq 0'$ so $b^{(\text{ht}(c))} \geq 0^{(\text{ht}(c)+1)} \not\leq c$. Thus $b \in \text{TS}(\pi^1)$, so $\alpha_{b,\xi} | \alpha_{b,v}$, a contradiction.

We now show that $\alpha_{a,v} \subseteq \alpha_{a,\kappa}$. We assume that there are $\pi^k \in \text{PL}(\sigma^m, \xi, v)$ and r such that $a \in \text{TS}(\pi^r)$ for some derivative or antiderivative π^r of π^k, in order to obtain a contradiction. Let the i-target sets for R_{π^k} be given as $\text{TS}(c, i)$, and let $n = \text{ht}(c)$. By Lemma 9.5.5 (Weight Comparison), if π is the initial derivative corresponding to the final entry of π^k into $\text{PL}(\sigma^m, \xi, v)$, then $\text{wt}(\text{up}(\pi^r), \text{init}(\text{up}(\pi^r), \pi)) \in \text{dom}(\alpha_{b,\xi})$. Now $r \leq \text{lev}(a)$. If $r < \text{lev}(b)$, then it follows from Lemma 10.2.3(iii) (Target Set) that $\alpha_{b,\xi}$ and $\alpha_{b,v}$ are incompatible; but as backtracking was implemented, $\alpha_{b,\xi} \subseteq \alpha_{b,v}$, yielding a contradiction.

Otherwise, we must have $r = \text{lev}(a) = \text{lev}(b)$. If $n+1 < r$, then $b \in \text{TS}(\pi^r)$, so $\alpha_{b,\xi} | \alpha_{b,v}$, yielding a contradiction. If $n + 1 = r$ and $b \in \text{TS}(\pi^r)$ then $b \leq c$; but $a \leq b$ and $\text{lev}(a) = r$, so $a \leq c$ and $a \in \text{TS}(\pi^r)$, again yielding a contradiction. Otherwise, we may assume that $n + 1 > r$ so $n \geq r$. As $n = \text{ht}(c)$ and $b \leq 0^{(r)}$, $b^{(n-r)} \leq 0^{(n)} \leq c$. But $b^{(n-r)} = (a')^{(n-r)} = a^{(n-r+1)}$, so $a \in \text{TS}(\pi^r)$, again yielding a contradiction. ⊣

We show that all requirements are satisfied.

LEMMA 10.4.2 (Comparability Lemma). *All type* 3 *requirements are satisfied.*

PROOF. Immediate from Lemma 10.2.3(i) (Target Set) and the definition of target sets for requirements. ⊣

LEMMA 10.4.3 (0-Satisfaction Lemma). *Every requirement of type* 0 *is satisfied.*

PROOF. Fix a requirement $R = R^{0,k}_{e,b,a}$ of type 0. By the standard initial assignment, $R = R_{\eta^\ell}$ for some $\eta^\ell \subset \Lambda^\ell$ such that η^ℓ is Λ^ℓ-free. Let ν^ℓ be the immediate successor of η^ℓ along Λ^ℓ. Let $\nu^i = \text{out}^i(\nu^\ell)$, $\eta^i = (\nu^i)^-$ for $i \leq \ell$, and $x = \text{wt}(\eta^k, (\text{init}(\eta^k, \eta)))$. As is our convention, we omit the superscript 0. Note that by Lemma 8.6.1(ii) (Freeness Transmission), η^j is Λ^j-free for all $j \leq \ell$, so by Lemma 8.3.7(ii) (Free and True), ν^j is Λ^j-true for all $j \leq \ell$. In fact, the same pattern will occur for any $\tau^\ell \in T^\ell$, so there are infinitely many $\tau \subset \Lambda^0$ such that $\tau^j = \lambda^j(\tau)$ is Λ^j-true for all $j \leq \ell$, and τ^- is Λ^0-free. We note that by Lemma 10.4.1 (Backtracking Implementation) and the construction, clauses (i)–(iii) of Definition 9.6.3 hold. Furthermore, by clause (i) of Definition 9.6.3, Lemma 9.5.4 (Similarity) for τ as above and the construction, if $\nu \subseteq \tau$, then $\alpha_{c,\nu} \subseteq \alpha_{a,\tau} \subset A_c$ for all $c \in P$, so clause (iv) of Definition 9.6.3 holds. Thus by Lemma 9.6.4 (Constraint), the successor of η along τ is not constrained, nor is any such τ.

First assume that Case 1 of the construction is followed. By Lemma 10.4.1 (Backtracking Implementation), a unique backtracking extender exists in this case, so η is that backtracking extender; furthermore, $\alpha_{a,\delta} \subseteq \alpha_{a,\eta}$. Then $\alpha_{b,\nu}(x) = 1$ and so $A_b(x) = 1$. By the conditions of Case 1, $\Phi_e(\alpha_{a,\delta}; x)[\text{wt}(\delta)] = 0$ and $\lambda^j(\mu) \subseteq \lambda^j(\delta)$ for all $j \leq \ell$. As $\alpha_{a,\delta} \subseteq \alpha_{a,\eta}$, $\Phi_e(\alpha_{a,\eta}; x)[\text{wt}(\eta)] = 0$. Now the only nodes switched by ν with weight in $\text{dom}(\alpha_{a,\delta})$ are the antiderivatives η^j of η^k, and $a \notin \text{TS}(\eta^j, j)$ for $j \in [k, \ell]$. Thus $\alpha_{a,\eta} \subseteq \alpha_{a,\nu}$ and so $\Phi_e(\alpha_{a,\nu}; x)[\text{wt}(\nu)] = 0$. As $\alpha_{a,\nu} \subset A_a$, $\Phi_e(A_a; x) = 0$, so R is satisfied.

Now suppose that Case 2 of the construction is followed. First assume that $A_b(x) = 1$. Then $\alpha_{b,\nu}(x) = 1$, and $\Phi_e(\alpha_{a,\eta}; x)[\text{wt}(\eta)] = 0$. Now $\alpha_{b,\eta}(x) = 1$, else η would require attention for $\xi = \eta$ and so its successor along Λ^0 would be constrained, which is not the case. Thus ν is non-switching, so $\alpha_{a,\eta} \subseteq \alpha_{a,\nu}$, implying that $\Phi_e(\alpha_{a,\nu}; x)[\text{wt}(\nu)] = 0$. As $\alpha_{a,\nu} \subset A_a$, $\Phi_e(A_a; x) = 0$, and R is satisfied.

Now assume that $A_b(x) = 0$. We assume that $\Phi_e(A_a; x) = 0$ and derive a contradiction, thereby showing that R is satisfied. As noted above, we can choose a $\tau \supset \nu$ sufficiently long so that $\lambda^j(\tau) \subset \Lambda^j$ for all $j \leq \ell$ and l^j_τ is greater than the use for $A^{[j]}$ of the computation $\Phi_e(A_a; x) = 0$. Thus $\alpha_{c,\nu} \subseteq \alpha_{a,\tau} \subset A_c$ for all $c \in P$, and so $\Phi_e(\alpha_{a,\tau}; x)[\text{wt}(\tau)] = 0$ and $\alpha_{b,\nu}(x) = \alpha_{b,\tau}(x) = 0$. We now see that τ requires attention for η, so τ is constrained, yielding the desired contradiction. ⊣

LEMMA 10.4.4 (Well-Definedness and Totality Lemma). *Every functional* $\Delta = \Delta_{b,a}^{1,m}(A_b)$ *introduced in the construction is well-defined and total.*

PROOF. We note that by the definition of the lengths of sections of the function α, if $\alpha_{a,\xi}$ and $\alpha_{a,\eta}$ are compatible, then they are comparable, and in that case, if $\xi \subset \eta$ then $\alpha_{a,\xi} \subseteq \alpha_{a,\eta}$. Thus by induction, if a new axiom $\Delta_\nu(\alpha_{b,\nu}; e) = r$ is to be declared as action for η, and an axiom $\Delta_\rho(\alpha_{b,\rho}; e) = 1$ has been declared earlier as action for ξ with $\alpha_{b,\rho} \subseteq \alpha_{b,\nu}$, then $\eta = \nu^-$, $\xi = \rho^-$, $\Phi_e(\alpha_{a,\xi}; e)\downarrow$ and $\alpha_{a,\xi} \subseteq \alpha_{a,\rho}$. As $a' = b$, $a \leq b$ so by Lemma 10.4.2 (Comparability), we must also have $\alpha_{a,\rho} \subseteq \alpha_{a,\nu}$. Thus $\Phi_e(\alpha_{a,\eta}; e)\downarrow$, and so the axioms declared have the same value.

Suppose that a new axiom $\Delta_\nu(\alpha_{b,\nu}; e) = r$ is to be declared as action for η, and an axiom $\Delta_\rho(\alpha_{b,\rho}; e) = 0$ has been declared earlier as action for ξ with $\alpha_{b,\rho} \subseteq \alpha_{b,\nu}$. If $\Phi_e(\alpha_{a,\eta}; e)\downarrow$, then we would initiate the backtracking process at η thereby constraining η, so no action to declare axioms would be taken at η. Thus we again see that the axioms declared have the same value.

By the assignment process and the construction, Δ will be total. ⊣

LEMMA 10.4.5 (1-Satisfaction Lemma). *Every requirement of type* 1 *is satisfied.*

PROOF. The proof is virtually identical to that of Lemma 10.4.3 (0-Satisfaction), so is not presented. One first needs to make the following modifications: type 0 is changed to type 1, x is changed to e, $\Phi_e(\alpha_{a,\zeta}; x)[\mathrm{wt}(\sigma)] = 0$ is changed to $\Phi_e(\alpha_{a,\zeta}; e)[\mathrm{wt}(\sigma)]\downarrow$ for any pair of nodes ζ and σ, and $\Phi_e(A_a; x) = 0$ is changed to $\Phi_e(A_a; e)\downarrow$. ⊣

We now turn our attention to the satisfaction of requirements of type 2. We define functionals $\Delta_{b,a}^{2,k}$ corresponding to requirements $R_{e,b,a}^{2,k}$ as follows. Fix $x, s \in \mathbb{N}$. If there do not exist $i \in (0, \ell]$, $\xi \in T^0$ and $\xi^i \in T^i$ such that $x = \mathrm{wt}(\xi^i, \xi)$, then we define $\Delta_{b,a}^{2,k}(X; x, s) = 0$ for all sets X. (Note that the set of such X is computable.) Otherwise, fix such ξ and i, and define $\Delta_{b,a}^{2,k}(\alpha_{a,\eta}; x, s) = m$ at stage η of the construction for all $s < \ell_\eta^i$ for which no axiom from an oracle $\alpha_{a,\rho}$ for some $\rho \subset \eta$ for which $\alpha_{a,\rho} \subseteq \alpha_{a,\eta}$ has previously been defined, setting $m = 1$ if $\alpha_{b,\eta}(x) = 1$, and $m = 0$ otherwise.

LEMMA 10.4.6 (2-Satisfaction Lemma). *All requirements of type* 2 *are satisfied.*

PROOF. Fix $x \in \mathbb{N}$. If there do not exist $j \in (0, \ell]$, $\eta \in T^0$ and $\eta^j \in T^j$ such that $x = \mathrm{wt}(\eta^j, \eta)$, then $\Delta(A_a; x, s) = 0 = A_b(x)$ for all s as required. Thus we may fix such j, η and η^j. By Lemma 10.2.3(ii) (Target Set), $a \in \mathrm{TS}(\xi^{j-1})$ for all $\xi^{j-1} \in T^{j-1}$ such that $\mathrm{up}(\xi^{j-1}) = \eta^j$. By Lemma 9.4.2 (Principal Switching), if $\sigma \subset \tau \subset \Lambda^0$, $\alpha_{b,\sigma}(x)\downarrow = 1$ $\alpha_{b,\tau}(x)\downarrow = 0$, ξ^{j-1} is the principal derivative of η^j along $\lambda^{j-1}(\sigma)$, $\xi = \mathrm{init}(\xi^{j-1}, \sigma)$ and $y = \mathrm{wt}(\xi^{j-1}, \xi)$, then

ξ^{j-1} will be switched between σ and τ, and at its last switch before each, its outcomes will be of different types. Thus $\alpha_{a,\sigma}(y)\downarrow = 0$ and $\alpha_{a,\tau}(y)\downarrow = 1$, so the oracles for the axioms declared at σ and τ will be incompatible. It now follows that $\lim_s \Delta(A_a; x, s) = A_b(x)$ as required. $\dashv$

Theorem 10.1.2 now follows from Lemma 10.2.8 (Requirement), Lemma 10.4.4 (Well-Definedness and Totality), and Lemmas 10.4.3, 10.4.5, and 10.4.6 (j-Satisfaction for $j \leq 2$), 10.4.2 (Comparability), and Lemma 9.7.3 (REA-ness). $\dashv$

Chapter 11

INFINITE SYSTEMS OF TREES

Some constructions will take place on an ω-sequence of trees. One example of such a construction is the construction of an *intermediate degree*, i.e., a degree $\boldsymbol{a}$ such that $\boldsymbol{0}^{(n)} < \boldsymbol{a}^{(n)} < \boldsymbol{0}^{(n+1)}$ for all $n \in \mathbb{N}$. Other examples are constructions to prove theorems about the arithmetical degrees. While we will not present constructions requiring infinitely many trees, we will briefly describe how to extend the framework to cover such constructions in this chapter.

11.1. Infinite Systems: A Brief Analysis

A typical construction will have an associated set of requirements effectively ordered into an ω-sequence $\langle R_i : i \in \mathbb{N}\rangle$. There will be a basic module M_i associated with each requirement R_i, and this basic module will have a *level* which will coincide with the largest level of any of the non-termial nodes of he module. The assignment of requirements to nodes of trees will proceed by induction. We will need both an *initial assignment algorithm* and a *derived assignment algorithm* to be specified for each construction.

We begin by assigning M_0 to the shortest node of each tree whose level is greater than or equal to the level, m_0, of M_0. For $k \geq m_0$, we will have associated a node of M_0 along an initial segment of each path through T^k, and each node of T^k to which a terminal node of M_0 is assigned will begin a new block. The derived assignment algorithm can be applied to T^j for $j < m_0$, and will determine the antiderivative operation along each path through T^j until we reach a first node ρ^j along the path such that a ρ^j-path through T^j is completed at $(\rho^j)^-$; all the nodes $\subset \rho^j$ will be antiderivatives of nodes of T^{m_0} to which this implementation of M_0 has been assigned. Thus when we reach a node $\rho^j \in T^j$ that begins a new block, it will be the case that $\lambda^{m_0}(\rho^j)$ will be a node that begins a new block of T^{m_0}.

We must now define the antiderivative function up going from T^k to T^{k+1} for $k \geq m_0$. We note that by (2.6.3) and Lemma 8.1.1(i), if η^k is a node in the initial block of T^k for $k \geq m_0$, then the same node of M_0 will be assigned to both η^k and $\lambda(\eta^k)$. Thus defining $\text{up}(\eta^k) = \lambda(\eta^k)$ for $k \geq m_0$ will produce the same assignment as if we had started on T^{k+1} and made the assignment via the derived assignment algorithm.

We will always have the induction hypothesis that if a ρ-block through T^0 is completed at $\tau \subset \rho$ with $\rho^- = \tau$, then for all k, a $\lambda^k(\rho)$-block through T^k is completed at $(\lambda^k(\rho))^-$, and requirements are assigned to all nodes $\subset \lambda^k(\rho)$. Furthermore, as only finitely many requirements have been assigned to this point, there will be a largest level i of requirements assigned to this point. We now follow the initial assignment process for T^k for $k \geq i$ and the derived assignment process for T^j for $j < i$, choosing a module M_i for T^i and proceeding as in the definition of the initial block. We leave it to the reader to show that these steps can consistently be carried out.

We note that all the lemmas about the framework proved in Chapters 8 and 9 hold with virtually identical proof. The only change that is required is a restatement of the Initial Assignment Property (Definition 2.7.2); instead of referring to the satisfaction of all requirements on a fixed tree T^n, we look at the collection of all paths $\Lambda^k \in [T^k]$ for $k \in \mathbb{N}$, and require that the satisfaction of all requirements that are assigned along some Λ^k to free nodes, where k is greater than or equal to the level of the module for the requirement, implies the theorem being proved.

REFERENCES

[1] C. J. Ash, *Stability of recursive structures in arithmetical degrees*, ***Annals of Pure and Applied Logic***, vol. 32 (1986), pp. 113–135.

[2] ———, *Labelling systems and r. e. structures*, ***Annals of Pure and Applied Logic***, vol. 47 (1990), pp. 99–119.

[3] S. B. Cooper, *Degrees of unsolvability complementary between recursively enumerable degrees, Part i*, ***Annals of Mathematical Logic***, vol. 4 (1972), pp. 31–73.

[4] R. M. Friedberg, *Two recursively enumerable sets of incomparable degrees of unsolvability*, ***Proceedings of the National Academy of Sciences of the United States of America***, vol. 43 (1957), pp. 236–238.

[5] S. C. Kleene, ***Introduction to Metamathematics***, Van Nostrand, New York, 1952.

[6] J. Knight, *A metatheorem for constructions by finitely many workers*, ***The Journal of Symbolic Logic***, vol. 55 (1990), pp. 787–804.

[7] A. H. Lachlan, *Lower bounds for pairs of recursively enumerable degrees*, ***Proceedings of the London Mathematical Society***, vol. 16 (1966), pp. 537–569.

[8] ———, *On some games which are relevant to the theory of recursively enumerable sets*, ***Annals of Mathematics* (2)**, vol. 91 (1970), pp. 291–310.

[9] S. Lempp, ***Lecture Notes on Priority Arguments***, to appear.

[10] S. Lempp and M. Lerman, *Priority arguments using iterated trees of strategies*, ***Proceedings of the Oberwolfach Conference on Recursion Theory, 1989*** (K. Ambos-Spies, G. H. Müller, and G. E. Sacks, editors), Lecture Notes in Mathematics, Springer-Verlag, Berlin, Heidelberg, New York, 1991, pp. 277–296.

[11] ———, *The existential theory of the poset of r.e. degrees with a predicate for single jump reducibility*, ***The Journal of Symbolic Logic***, vol. 57 (1992), pp. 1120–1130.

[12] ———, *A general framework for priority arguments*, ***The Bulletin of Symbolic Logic***, vol. 1 (1995), pp. 189–201.

[13] ———, *The decidability of the existential theory of the poset of recursively enumerable degrees with jump relations*, ***Advances in Mathematics***, vol. 120 (1996), pp. 1–142.

[14] ———, *Iterated trees of strategies and priority arguments*, ***Archive for Mathematical Logic***, vol. 36 (1997), pp. 297–312.

[15] S. LEMPP, M. LERMAN, and F. WEBER, *Minimal pair constructions and iterated trees of strategies*, ***Logical Methods: In Honor of Anil Nerode's 60th Birthday*** (J. Crossley, J. Remmel, R. Shore, and M. Sweedler, editors), Birkhäuser, Boston, Basel, Berlin, 1993, pp. 512–524.

[16] M. LERMAN, *The existential theory of the poset of Turing degrees with least element and jump is decidable*, to appear.

[17] M. LERMAN, R. A. SHORE, and R. I. SOARE, *The elementary theory of the recursively enumerable degrees is not $\aleph_0$-categorical*, ***Advances in Mathematics***, vol. 53 (1984), pp. 301–320.

[18] G. MICHALSKI, ***On Foundations of Recursion Theory***, Ph.D. thesis, University of Notre Dame, 1996.

[19] A. A. MUČNIK, *On the unsolvability of the problem of reducibility in the theory of algorithms*, ***Doklady Akademii Nauk SSSR, N. S.***, vol. 108 (1956), pp. 194–197, (Russian).

[20] E. L. POST, *Recursively enumerable sets of positive integers and their decision problems*, ***Bulletin of the American Mathematical Society***, vol. 50 (1944), pp. 284–316.

[21] G. E. SACKS, ***Degrees of Unsolvability***, Annals of Mathematical Studies, vol. 55, Princeton University Press, Princeton, NJ, 1963.

[22] ———, *Recursive enumerability and the jump operator*, ***Transactions of the Mathematical Society***, vol. 108 (1963), pp. 223–239.

[23] J. R. SHOENFIELD, *On degrees of unsolvability*, ***Annals of Mathematics***, vol. 69 (1959), pp. 644–653.

[24] ———, *Applications of model theory to degrees of unsolvability*, ***Symposium on the Theory of Models*** (J. W. Addison, L. Henkin, and A. Tarski, editors), North-Holland, Amsterdam, 1965, pp. 359–363.

[25] R. I. SOARE, ***Recursively Enumerable Sets and Degrees***, Perspectives in Mathematical Logic, Springer-Verlag, Berlin, Heidelberg, New York, 1987.

[26] S. K. THOMASON, *Sublattices of the recursively enumerable degrees*, ***Zeitschrift für Mathematische Logik und Grundlagen der Mathematik***, vol. 17 (1971), pp. 273–280.

[27] C. E. M. YATES, *A minimal pair of recursively enumerable degrees*, ***The Journal of Symbolic Logic***, vol. 31 (1966), pp. 159–168.